Mathematics Study Resources

Volume 11

Series Editors

Kolja Knauer, Departament de Matemàtiques Informàtic, Universitat de Barcelona, Barcelona, Spain

Elijah Liflyand, Department of Mathematics, Bar-Ilan University, Ramat-Gan, Israel

This series comprises direct translations of successful foreign language titles, especially from the German language.

Powered by advances in automated translation, these books draw on global teaching excellence to provide students and lecturers with diverse materials for teaching and study.

Stefan Müller-Stach

The Code of Mathematics

Proof and Truth

 Springer

Stefan Müller-Stach
Universität Mainz
Mainz, Germany

ISSN 2731-3824   ISSN 2731-3832   (electronic)
Mathematics Study Resources
ISBN 978-3-662-69482-4   ISBN 978-3-662-69483-1   (eBook)
https://doi.org/10.1007/978-3-662-69483-1

Translation from the German language edition: "Der Code der Mathematik" by Stefan Müller-Stach, ©
Der/die Herausgeber bzw. der/die Autor(en), exklusiv lizenziert an Springer-Verlag GmbH, DE, ein Teil
von Springer Nature 2023. Published by Springer Berlin Heidelberg. All Rights Reserved.

This book is a translation of the original German edition "Der Code der Mathematik" by Stefan Müller-
Stach, published by Springer-Verlag GmbH, DE in 2023. The translation was done with the help of an
artificial intelligence machine translation tool. A subsequent human revision was done primarily in terms
of content, so that the book will read stylistically differently from a conventional translation. Springer
Nature works continuously to further the development of tools for the production of books and on the
related technologies to support the authors.

This Springer imprint is published by the registered company Springer-Verlag GmbH, DE, part of Springer
Nature.
The registered company address is: Heidelberger Platz 3, 14197 Berlin, Germany

If disposing of this product, please recycle the paper.

Preface

La musique est une mathématique mystérieuse dont les éléments participent de l'infini.

—Claude Debussy (Musica, May 1903).

Our Earth is part of a gigantic universe, the fundamental principles of which are explained by physics. Despite all the mysteries it still holds, we know a great deal about it. In mathematics, there are universes of a completely different nature. They form infinite reservoirs of objects that are not found in reality. Together with the precise mathematical calculus, they unfold from them a world of fascinating beauty and diversity.

The connection between mathematics and physics goes much deeper than this analogy. Numerous sciences cannot express their theories without the aid of mathematical structures and mathematical calculus forms the basis of digitisation and other technologies. Galileo Galilei once called mathematics the language of nature. Eugene Wigner spoke of the

Unreasonable effectiveness of mathematics in the natural sciences.[1]

It was a dream far beyond that of Gottfried Wilhelm Leibniz to construct a universal scientific language—the Lingua universalis—which generalises mathematics. As Umberto Eco and others have noted, this idea has never been fully realised. Its systematic limits remained—despite important milestones in computer science—unclear.[2]

In May 2018, I gave a lecture titled Truth, proof, thought, identity at the Mainz Studium Generale. In it, I explained the basics of homotopy type theory, which builds on dependent type theory and can better grasp the concepts of equality, isomorphism and equivalence through built-in topological concepts than the traditional approach via set theory. Such ideas are related to fundamental questions of computer science, philosophy and physics.

Since type theory resembles the code of a programming language, this approach enables the machine-checkable verification of proofs and algorithms. This development has the potential to possibly change the research, learning and publication culture far beyond mathematics, as Leibniz had envisaged.

This book presents a detailed version of the lecture for an interested audience. To achieve good access, a natural structure of mathematics is outlined, which deviates from the usual curriculum. Since in particular the later chapters have a higher level of difficulty, numerous notes and references to further literature should facilitate reading and compensate for the missing details in the popular scientific presentation. The selected aspects span a historical arc from antiquity to the present day. This once again illustrates that scientific knowledge gain usually takes place on long time scales.

This text is based on a machine English translation of the German first edition. The postediting work and all illustrations were done by myself. German quotes and titles of publications in the main text have been translated with references pointing to the original source. I would like to thank the following people for remarks: Lambert Alff, Nikoo Azarm, Carl–Heinz Barner, Manuel Blickle, Elke Brendel, Annika Denkert, Petra Gerster, Jürgen Jost, Thomas Metzinger, Sieglinde Müller–Stach, Verena Nörthen, Andreas Rödder, Silvana Rödder, Andreas Rüdinger, Peter Stoll, Thomas Streicher, Christian Tapp, Ulrich Volp and Rainer Wieland.

Mainz and Neustadt–Haardt, Germany   Stefan Müller-Stach
March 2024

About This Book

As a guide, we want to anticipate some themes and questions that we will encounter. Two fundamental questions are:

What is a thought?

and

Where are the abstract concepts of our thinking located?

Despite numerous insights from philosophy and the life sciences, there are still no simple answers to these two questions about the nature of thoughts and the location of our thinking and consciousness. Leibniz pointed out in his mill example[3] that any journey through our brain would only reveal a machine—a mill from the inside—without perceptions, consciousness or other qualia and emergent phenomena of life being recognisable.

In the philosophy of Plato and its tradition, abstract objects and concepts were located in a world of ideas outside of physical reality. The Platonic world of ideas has a natural counter-position in nominalism. An introductory question to this complex of topics is:

What is a number?

Numbers are universally known objects that do not exist in reality themselves, but only in the form of counts. We handle small numbers confidently, but very large numbers completely elude our imagination. The same applies to ideal geometric shapes, such as circles, which exist in their mathematical purity only approximately in reality. Such figures are special cases of a general concept of space. Algebraic structures like numbers can be assigned suitable spaces, so that we can consider the concept of number as a part of the concept of space. Thus, the nature of numbers and geometric objects is naturally connected with the following question:

What is the most general concept of space?

We approach this question by dealing in detail with examples of topological spaces and their generalisations.

Since we are already deeply involved in mathematics with such discussions, we deal with its working methods and its influence:

What is the working method of mathematics?

and

What role does it play in our culture?

In doing so, we exemplarily address some basic mathematical concepts and describe various useful algorithms and exciting applications.

The foundation of mathematics is its syntax. This includes underlying deductive systems, in which proofs can be conducted using a formal language. The following question can be asked:

Is provability a kind of computability?

This is indeed correct. Every proof is a computation in a deductive system. In a certain sense, conversely, every computation is a proof for the statement that asserts the result of the computation. The question can therefore be answered in the affirmative in a general sense. A more difficult question is the decision problem (German Entscheidungsproblem) of Hilbert and Ackermann:

Can it be decided which propositions in an axiomatic theory are provable by deductive methods?

Propositions correspond to formulas without free variables in the formal language of a deductive system. Sometimes they are also called sentences or theorems and often we refer to them informally as statements. A decision would be an algorithm that always terminates and delivers the result 1 if the proposition is provable and 0 otherwise. However, the answer to this question is a clear no, as Alan Turing and Alonzo Church showed in their famous works[4] of 1936. In addition to Emil Post and Stephen Kleene, they provided a precise concept of computability and used a trick by Gödel, in which mathematical propositions or Turing machines are assigned a natural number as Gödel number. This leads to the decision problem and other undecidable problems being reduced to a yes/no decision for natural numbers, which can be technically reduced to the undecidability of the halting problem for Turing machines.

An open question is whether there is a concept of computability beyond Turing computability, often called hypercomputing, and whether the human mind is superior to a computer or not:

What distinguishes human intelligence from a computer?

Alan Turing himself thought intensively about such questions and developed the Turing test[5] in this context.

Part of the book is dedicated to the concepts of truth and semantics:

How can the concept of truth be defined?

and

What role does semantics play in mathematics?

The first question was already asked in antiquity by Aristotle and others and has led over many centuries to a multitude of attempts at explanation, of which the correspondence theory of truth was the most widespread.

A possible verification of the concept of truth is aimed at by Leibniz's search for a universal scientific language, the Lingua universalis. In such a language, the verification of the truth of statements in a calculus adapted to the subject of investigation would be possible by proof in a syntactic way. This approach is fundamentally pragmatic, as the concept of provability in deductive systems is well understood. The ideas of Leibniz led to the emergence of mathematical logic.

Building on this, Alfred Tarski provided a language-analytical definition of truth Although Tarski claims to want to further develop the correspondence theory of truth[6] in the 20th century, which works well at least for statements in deductive systems with formal object languages L, as they occur, for example, in mathematics. The self-referential everyday language of daily life, which carries its truth predicate within itself, does not meet these conditions due to the existence of self-referential antinomies. Tarski used a metalanguage M, which is richer than L, to realise a truth predicate for statements in L within M. The chosen mathematical model associated with M is referred to as the semantics of L and the possible transitions from L to M are referred to as interpretations. In many cases, M is given by the metalanguage of the axiomatic set theory with the Zermelo–Fraenkel axioms and possibly additional axioms about Grothendieck universes.

The most important example of Tarski's method is the formal object language L_{ar} of Dedekind–Peano arithmetic. In addition to the set-theoretic interpretation in the standard model

$$\mathbb{N} = \{0, 1, 2, \dots\},$$

L_{ar} allows various interpretations in non-standard models with differing properties. A statement in L_{ar} is called true if its interpretation in the respective set-theoretic model is fulfilled. This is proven with the help of the axioms of set theory. Tarski has shown that the truth predicate for Dedekind–Peano arithmetic is not definable within L_{ar}. In his theorem on the undefinability of truth, he even proved that the Gödel numbers of the true theorems in the standard model form an undecidable set.

With Tarski's concept of truth, we can approach a new question:

Is truth the same as provability?

Since the truth of a statement in L is defined by provability in a selected interpretation with the help of a metalanguage M, this question seems to be clarified in a certain way, albeit not within L. Indeed, there are statements in certain deductive systems L that are neither provable nor refutable in L and whose validity can only be clarified in a selected metalanguage. This was proven by Kurt Gödel in his famous incompleteness theorem.[7] The continuum hypothesis falls into a similar category, as it cannot be decided with the usual Zermelo–Fraenkel axioms of set theory. Such statements are called undecidable, even though the concept of undecidability is used in a different way than in the decision problem. The answer to our question is thus:

Truth depends on an interpretation. The proof is carried out in the corresponding metalanguage.

Because of these observations, it makes sense to consider hierarchies of extensions of formal languages.

Mathematical theories are usually formulated in the language of predicate logic and set theory. However, this is not the only possibility. The search for alternatives to traditional set theory has been approached on the one hand in the form of (dependent) type theory through the work of the logician Per Martin–Löf and on the other hand through higher categories and infinity categories, building on ideas from Alexander Grothendieck.[8] This leads us to an overarching theme:

Which form of the foundations of mathematics is most suitable for structuralist thinking?

By foundations of mathematics, we understand thought systems that can describe large parts of mathematics. Vladimir Voevodsky and others have significantly further developed the theory of Martin–Löf. This research direction, also known as homotopy type theory, has incorporated topological and homotopy theoretical aspects into its basic concepts. Through this artifice, type theory has become a mature theory and can handle the concept of equality and its variants isomorphism, symmetry and equivalence more adequately than, for example, set theory. The common concept of equality has proven to be too rigid over time. The question behind this is:

What does the equation $A = A$ mean?

This question, which is also significant in philosophy, leads in type theory to the definition of an identity type, which is the starting point for new forms of equality.

Dependent type theory forms through its deductive system the basis of new software tools, which allow the provable verification of proofs and algorithms and in the future will enable both powerful assistance systems for research and teaching as well as likely change publication practices. Large parts of mathematics have already been formalised in such systems.

Through type theory and category theory, new concepts emerge for the foundations of mathematics. Often a given dependent type theory is viewed as an object language and a (higher) categorical semantic interpretation is considered. Through such interpretations, the traditional mixing of syntax and semantics in set theory can be resolved. Going even further, these three foundations of mathematics can each be viewed as an object language or metalanguage and are mutually interpretable. From such an overarching perspective, a certain disenchantment of the concept of semantics results, which is reduced to the syntax of a chosen metalanguage. On the other hand, the mathematical concept of truth thereby receives a solid foundation that can no longer be relativised.

As a result of our considerations, a new perspective on mathematics emerges, which allows a more structuralist understanding than before, opened up and significantly expanded the possibilities for the verification of proofs and algorithms. This development goes beyond mathematics and makes a relevant contribution to Leibniz's goal of the Lingua universalis.

Contents

What is mathematics and what are its subjects? This question is not easy to answer. From observing the world, we have always drawn inspiration for mathematical concepts. Essentially, however, it is an a priori science, as it is based neither on experience nor on other assumptions. Its simplest features seem to be evolutionarily ingrained in our brain.

Based on logic,[9] which has served as a tool for proofs since antiquity, mathematical concepts are formulated in deductive systems. Their formal languages must be rich enough to describe all mathematical objects that underlie the individual concepts being considered. Such systems are called foundations of mathematics. Axiomatic set theory is the most widespread form among them. Two other forms of foundations are category theory and type theory.

When the first arithmetic and geometric concepts emerged in our cultures, the question already arose as to what form of existence the underlying objects possess and in what way they can be identified with each other. The Platonic world of ideas postulated an existence at a specially designated place. It is closely related to the concept of equality among abstract objects. The aim of better understanding the role of different forms of equality and even more general equivalence concepts has triggered promising developments in research in recent mathematics.

Mathematical Objects and Their Identification

We will first consider abstract objects and illustrate the mathematical way of thinking using fundamental arithmetic and geometric concepts to lead from there to deeper questions about the foundations of mathematics.

If you believe you know what concept the natural numbers

$$0, 1, 2, 3, 4, 5, \ldots$$

and their arithmetic as a whole form, what the concept of a number is or what kind of object the single number

$$5$$

represents, then you may be mistaken. Even mathematically trained people have always had difficulty explaining this. It was not until the 19th century that Georg Cantor, Richard Dedekind and Gottlob Frege began to define the natural numbers precisely and to prove their desired properties.

The notation for the number 5 varied in different world languages and throughout history. This is not a problem, because we all—even small children, some animal species and people in cultures without significant schooling—have a sense for small numbers and elementary mathematical problems.[10] Any intelligent extraterrestrial being that would visit us would probably also have a concept of it, because it would be able to make a connection when looking at our hands with their 5 fingers and would probably have a designation for this number, depending on what form of communication it uses. But is this the same idea for all people and are all avatars of the number 5 somehow the same? In other words, does the equation

$$5 = 5$$

always hold? Our intuition tells us that this is correct and all variants of the number 5 can be identified. This impression probably comes from the fact that we all mean the same thing in our daily dealings with it and can communicate perfectly well about small numbers. Studies in early childhood suggest that a sense for small numbers of objects is innate or can at least be learned quickly, similar to language acquisition. However, we do not understand why these facilities are present in the brain and how they are laid out there.

To make matters worse, there are complicated additive and multiplicative relationships between the numbers that occur within arithmetic. The equations

$$5 = 1 + 1 + 1 + 1 + 1$$

or

$$5 = 3 + 2$$

are to be regarded as non-trivial statements about the number 5. Here, a big problem apparently arises, namely the question of the identifiability of mathematical objects. The terms equality, identity, isomorphism and equivalence are commonly used in mathematics for this. The equation

$$5 = 5$$

and especially the different representations of the number 5 as sums of other smaller numbers touch deeper levels of mathematics. Richard Dedekind and Gottlob Frege recognised that natural numbers have infinitely many different realisations and equations like $5 = 3 + 2$ are in reality subtle assertions of equivalences between different objects, which need to be mathematically explained and proven. Frege attempted to solve the definition of natural numbers through his method of abstraction. Georg Cantor developed further concepts of transfinite numbers.[11]

The strongest form of equality, i.e., sameness, such as $5 = 5$, we will call definitional equality. Other forms of equality, such as $5 = 3 + 2$, are called propositional equality. Such different variants of equality become a subject of research in modern mathematics. This leads—besides a more precise definition of the concept of equality—to a better understanding of generalisations such as the concepts of isomorphism, equivalence and univalence.

In everyday life and in other sciences, the concept of equality is also important and has very pragmatic aspects. However, there are also gradations there. One aspect of this is the recognisability of objects and people. In particular, the word identity, and less often the word equality, is used in connection with the identification of people by means of IDs or fingerprints. A person changes a lot in life. Parts of our bodies have a shorter lifespan than we do. When we go to the hairdresser, or lose some hair or skin cells in some other way, no one will doubt that we have remained equal as a person and have retained our identity, for a human being is more than a collection of individual objects. In addition to the "external", a person changes over time also as a personality and is longer the same person in characteristics and appearance. After abrupt changes, the externally identical person may even no longer be equal as a person. The concept of equality in humans thus also touches on the psychological and social aspects of life and is, as in mathematics, to be distinguished from the concept of the same, which means complete agreement.

A composed piece of music which is provided with a fixed score will be played slightly differently at each performance. Nevertheless, it is equal as a piece of music that is being listened to. So there is a kind of concept of equality for pieces of music and musical performances which is not based on complete agreement. This is even more liberal in jazz compositions, where improvisation is done over given passages.[12] Here too, we never hear the same performance, but the concept of equality is more generously defined. Everyone immediately understands this example and recognises the same pieces of music without any problems. The same applies to other works of art. Every book, every piece of music and picture, which develops in its creation, remains equal from a certain stage, even if certain parts are still changed. Once an identity of the artwork has emerged, it remains preserved in the future.

The Concept of Equality in Frege and Leibniz

Frege dealt extensively with the concept of equality in his influential essay "On sense and meaning".[13] In it, he found a wonderful geometric example. He used triangle geometry and considered the three bisectors in a triangle. Because of a well-known theorem of triangle geometry, the intersection points of two bisectors in the triangle coincide. Thus, the equation $A = B$ results (see figure).

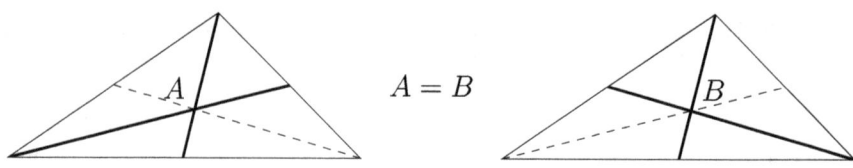

The Fregean example.

The proof for this arises from the fact that the three medians of a triangle always intersect at a common point, namely the centroid. Both A and B are therefore identical with this point and thus A and B are identical themselves. The equation $A = B$ therefore expresses the correctness of the geometric theorem of the centroid and is not a random identity of two a priori different points.

By the way, you may have noticed that in the above proof the two triangles in the figure are supposed to symbolise the same triangle, although they are only equal. To better see the respective intersection points, two identical copies of the triangle were drawn, a common trick. Our thinking thus uses variants of the concept of equality without hesitation.

In the same essay, Frege pointed out that in this example the two points A and B coincide and thus have the same meaning (i.e., reference or denotation), but that the respective sense, which lies in the definition of the points, is different. The difference between sense and meaning of signs, in his opinion, is expressed by the equation[14]

$$A = B.$$

In his text, Frege gives further examples of this. One of them comes from the two designations morning star and evening star for the planet Venus. The equation

$$morning\ star\ =\ evening\ star$$

expresses the coinciding meaning of both as planet Venus, while the sense on both sides of the equation differs by the time of day during observation. This distinction leads to a differentiation of the concept of equality in sense and meaning, or equivalently in intensional and extensional ways. Such investigations form the beginning of a philosophy of language with Frege, which in the course of his text leads to the consideration of thoughts and statements and their truth.

Even before Frege, Gottfried Wilhelm Leibniz introduced a concept of equality that still plays an important role in mathematics and elsewhere today. In his own words, he said:

Eadem sunt quorum unum potest substitui alteri salva veritate.[15]

Translated into today's language, the equality $A = B$ applies to two objects if in all investigations the object A can be replaced by B while maintaining the truth. This is therefore a substitutional concept of equality, which is weaker than the concept of equality of the indistinguishability and additionally presupposes a concept of truth. It becomes a rule of inference in formal mathematical calculi—called Leibniz's invariance rule—in the sense that identical arguments in relations are interchangeable. This will encounter us later in our considerations of type theory by Per Martin–Löf.

Leibniz also defined a broader concept of similarity or equivalence, which was extremely clear and forward-looking for its time. He distinguished between the similarity and congruence of two geometric figures:

Similia sunt quae singulatim discerni non possunt.[16]

Translated into modern language, two figures are therefore similar if they are indistinguishable when viewed separately. Leibniz thus differentiated between the size and shape of a figure. When two figures are viewed separately, their size recedes into the background and only their shape remains. This means that in geometric figures there is not only the strong concept of equality, i.e., congruence, but also the weaker concept of similarity or equivalence, which is just as significant. This example shows that Leibniz played a role in the discovery of the concept of equivalence.[17]

The examples discussed so far are very illustrative and unproblematic. So what is the actual problem of equality or equivalence? Mathematical objects such as vector spaces, groups or manifolds are not unique. Isomorphisms and equivalences identify them in mathematics and in applications in the form of structure-preserving transformations. For example, spacetime theories in physics are modelled on manifolds, but the physical reality is of course independent of the chosen modelling. The fundamental problem is that it is not easy to decide for two given objects whether they are isomorphic or equivalent. A well-known example of this is the homeomorphism problem for two topological spaces, which is undecidable in the sense of computability theory. Even if isomorphisms and equivalences from previous considerations are available, they may not be able to be specified concretely or this information may be lost in the course of proofs due to lack of suitable bookkeeping. Recent mathematical research is trying to remedy such difficulties.

The Concept of Equality in Philosophy

We want to take a small detour into the field of philosophy. The problem of equality is not only a question of mathematics, but also a deep problem of philosophy. Martin Heidegger gave a lecture titled "The theorem of the identity" in Freiburg (June 27th 1957), in which the question of identity is brought into connection with existence, i.e., being[18]:

The principle of identity is commonly formulated as: $A = A$. The principle is considered the supreme law of thought. We try to think about this principle for a while. Because we want to learn from the principle what identity is ... What the principle of identity, heard from its basic tone, states, is exactly what all Western thinking thinks, namely this: The unity of identity forms a basic feature in the being of beings.[19]

This essay by Heidegger is not easy to understand and delves deeply into the world of metaphysics. At the end of his text, Heidegger refers to Parmenides and quotes him with the words:

The same is thinking as well as being.[20]

So, Heidegger is concerned with the identity of the human being with his being and thinking. Behind the concepts of being and self is the question of the nature of human consciousness.

There is a remarkable final remark in this text which classifies computability in contrast to thinking:

Today, the thinking machine calculates thousands of relationships in a second. They are essenceless despite their usefulness.[21]

Behind this is the question of whether human thinking fundamentally differs from computability. Heidegger was convinced that our thinking has a completely different essence than the calculations of the then emerging computers.

In addition to Heidegger and Leibniz, Johann Gottlieb Fichte dealt with the question of equality. He pointed out the deep content of the equation $A = A$ and emphasised, similar to Henri Poincaré, that this statement presupposes a judgement of a subject.[22]

The English philosophers John Locke and David Hume also thought about equality. Thus, Locke writes in his book "An essay concerning humane understanding" from 1690:

Another occasion the mind often takes of comparing, is the very being of things, when, considering anything as existing at any determined time and place, we compare it with itself existing at another time, and thereon form the ideas of identity and diversity.[23]

Hume encountered similar questions in his essay "Of personal identity" from 1739:

There are some philosophers, who imagine we are every moment intimately conscious of what we call our self; that we feel its existence and its continuance in existence; and are certain, beyond the evidence of a demonstration, both of its perfect identity and simplicity ... Unluckily all these positive assertions are contrary to that very experience, which is pleaded from them, nor have we any idea of self after the manner it is here explained ... The mind is a kind of theatre, where several perceptions successively make their appearance; pass, re-pass, glide away, and mingle in an infinite variety of postures and situations. There is properly no simplicity in it at one time, nor identity in different; whatever natural propension we may have to imagine that simplicity and identity. The comparison of the theatre must not mislead us. These are only the successive perceptions that constitute our mind; nor do we have the most distant notion of the place where these scenes are represented, or of the materials of which it is composed ... We have a distinct idea of an object that remains invariable and uninterrupted through a supposed variation of time; and this idea we call that of identity or sameness ... that all the nice and subtle questions concerning personal identity can never possibly be decided, and are to be regarded rather as grammatical than as philosophical difficulties. Identity depends on the relations of ideas; and these relations produce identity.[24]

These sentences show what a critical mind David Hume was. He examined concepts in the finest detail and made detailed judgements, which were always committed to scepticism. Thus, the first sentence questions the whole idea of consciousness and separates it from the self of humans. Hume noted that man's assessment of his self might be more of a narrative. Indeed, the puzzle of the self and consciousness remains unsolved to this day.

Hume contributed greatly to the philosophy of the Enlightenment and spent a long time in the Parisian salon of Baron Paul Henri Thiry d'Holbach, the son of a wine-growing family from Edesheim in the German region of Palatinate. He brought his philosophical views into the discourse and had close personal contact with the people present there. Among them was the unique Denis Diderot, who, together with the mathematician Jean d'Alembert, was at that time editor of the famous Encyclopedia. This salon[25] was, like the Encyclopedia, committed to the spirit of the Enlightenment. Hume will encounter us again in connection with the concept of numbers by Frege.

The Platonic World of Ideas

Let's assume for a moment that we can identify all avatars of the number 5. Where then does this one idea of the number 5 live, if there is only one? Starting with Plato, many people, especially Leibniz and Frege, have asked similar questions also for geometric shapes.

Circle figure.

Consider a circle. With the depiction of geometric figures, different questions and problems are associated. A real circle would basically have no extension and zooming in should not produce a pixelated circle ring as in this illustration. Moreover, a real circle would actually be invisible. It can therefore be said with justification that a real circle has never been drawn or seen anywhere in the world. What we draw or see is always just an image of our mental representation of an ideal circle. All these approximations of a circle seem to us equivalent to this ideal representation.

The non-existence of numbers and real circles in reality brings us to the question of where such abstract objects can be located. This leads us to the Platonic world of ideas, also called Platonism or Platonic realism.[26] It attributes a kind of existence, i.e., a being, outside of our thoughts and outside of reality to abstract concepts such as numbers or geometric objects as well as the theorems that apply under these objects. The Platonic view goes back to Plato, a student of Socrates, who liked to record his thoughts in dialogues. Plato and his school considered this world of ideas—in contrast to Aristotle—as the superior world compared to physical reality. What kind of existence lies behind the Platonic world of ideas remains unclear from the surviving texts and it seems that Plato occasionally adjusted his position on this point throughout his life.

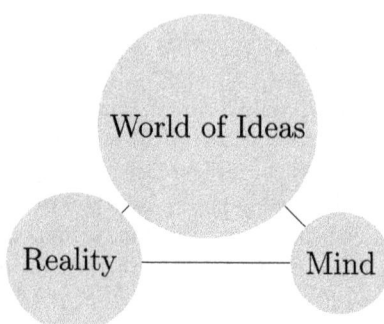

A picture of the Platonic world of ideas.

To explain the Platonic world of ideas, a three-part image is suitable (see figure). This model serves the careful mental separation of the three places reality, world of ideas and mind. The mind combines the possible human thoughts into one place. There are direct connections between mind, reality and ideas.[27]

Based on this image, there are several attitudes towards these three places. No one will doubt that our mind exists, because otherwise our thinking would not take place. The concept of reality is different. The immaterialism of George Berkeley claims that the material reality has no own existence and the world is only perceptible through perception.[28]

If we accept mind and reality, one attitude can be that the world of ideas is part of its own third place and strictly distinguishes itself from reality and mind. This picture is exactly the Platonic world of ideas. Secondly, there is the counterposition that there is no own place for the ideas and they are either part of reality or our thoughts. This view is called nominalism.[29] It means, that ideas appear in two forms, either in the form of natural laws and facts in physical reality or as human or divine thoughts.[30]

In mathematics, the Platonic view is widely held. This may be because mathematical theorems do not seem to be tautologies, but precious treasures that must be discovered. Indeed, many theorems in mathematics have a surprising statement and their proofs are hard to find. Examples of such theorems include Wiles' theorem or the 4-squares theorem. When they are found, or rather rediscovered, it feels like the discovery of a continent or the ascent of a high mountain.

Abstract thinking takes place in nominalism with signs and words and therefore with names, which explains the designation of this current in the philosophy of mathematics. This view already played a role in medieval thinking with Roscellinus or William of Ockham. The economy of thought, expressed by the principle of Ockham's razor, suggests a preference for nominalism, rejecting the unnecessary realism of objects in the Platonic world of ideas, which is also called Plato's beard as a counter-design to Ockham's razor.

The question of whether the Platonic or the nominalistic view is to be preferred is related to the dispute over universals[31] in philosophy. This is usually only applied to universals, i.e., to universal superordinate terms that name totalities of similar objects. A well-known theological example is the Trinity. A mathematical example of universals comes from Frege. He defined numbers like 5 as the class of all finite collections consisting of 5 elements. More precisely, he identified all sets with 5 elements to a new object that represents this class. However, this approach got him into trouble through Russell's paradox, as we will see. It is astonishing that the dispute over universals, the Platonic world of ideas and nominalism not only appeared in medieval scholasticism, but played a role throughout church history. Anselm of Canterbury, who co-founded scholasticism in the 11th century and whose name stands for an ontological proof of God, was a representative of the Platonic ideas. For him, it was plausible that the existence of a divine being could be compared with the existence of abstract objects or universals. His proof of God was based on the maximality of God's positive properties, which would not be fulfilled without his existence. Kurt Gödel, who confessed to the Platonic ideas, sketched in his last years an ontological proof of God with the help of a suitable modal logic, which is similar to that of Canterbury in the assumption of maximality.[32]

Doubts About the Platonic World of Ideas

No one can decide whether the Platonic world of ideas or nominalism is the correct standpoint. This plays only a subordinate role in our thinking about abstract objects using formal languages. In addition to the use of words, further symbols are necessary for this purpose. In his "Dialogus" from 1677, Leibniz wrote fittingly:

Cogitationes fieri possunt sine vocabulis ... At non sine aliis signis.[33]

Doubts about the Platonic world of ideas stem from the unresolved question of what kind of existence is attributed to ideas at a fictitious third place. For most people, it is clear that material objects have an existence in reality, i.e., in physical reality. In philosophy, even this is not self-evident and has led to the question of realism,[34] which asks about the existence of an external world independent of our consciousness. There are different and contradictory attitudes to this. The sceptical view of antirealism considers the world around us as a kind of fiction that arises in our consciousness as an internal construction. Realism, with its different manifestations, offers a counter-position. In this discourse, it remains unanswered in what sense abstract objects and ideas—beyond the question of realism—can exist in a Platonic sense.[35]

On the other hand, there are problems with the uniqueness of abstract objects. We have seen that it is inevitable that there are (infinitely) many avatars of the number 5. Paul Benacerraf and others have seen in this fact a contradiction to the Platonic world of ideas and formulated the Benacerrafian dilemma. It states that mathematics cannot simultaneously have a good semantics and a suitable ontology. Benacerraf's argument is based on the fact that any semantic explanation of the natural numbers necessarily produces (infinitely) many different realisations of each individual number, none of which is preferred, which contradicts the principle of Platonism.[36]

An alternative to the assumption of uniqueness in the Platonic world of ideas would be branched Platonic parallel worlds in the sense of possible worlds according to Leibniz, which house the respective avatars of numbers, geometric figures and other objects. In the further course, we will get to know new structuralist mathematical approaches that allow us to identify objects with the same structure. This approach defuses the dilemma of Benacerraf, but does not yet justify the Platonic world of ideas in a satisfactory way.

The Language of Set Theory

An informal concept of collections of objects has always been present in human minds. In 1851, Bernhard Bolzano described a naive concept of sets in his book "Paradoxes of the infinite"[37] which was common in his time. Between 1872 and 1882, Georg Cantor and Richard Dedekind exchanged numerous letters. Both had independently developed the basic concepts of set theory in a rigorous manner and openly shared their knowledge with each other.[38]

Dedekind developed set theory in his book "What are and what should the numbers be?"[39] as part of his investigations into natural numbers. Cantor had already needed aspects of set theory in his previous investigations into trigonometric series. Later, he wrote a series of works titled "On infinite linear point-manifolds".[40] As David Hilbert once said, Cantor had created a paradise within mathematics with the discovery of the world of infinite sets and their ascending transfinite hierarchy.

Both used—like Bolzano—a concept of sets in which sets were thought of as collections of arbitrary elements. This idea is often referred to today as naive or material set theory, because in it sets are seen in such a way that all elements $x \in A$ of a set A are known and it can be checked for each element x whether x is an element of A. Two sets are equal if they have the same elements. An important feature of this concept is the fact that an element x can occur in several sets. Thus, a naive set of the form

$$A = \{\text{Jan's dog, Heike's cat}\}$$

can exist, where the description stands for two specific animals and their owners. Set theory has various operations, such as the mappings

$$f : A \longrightarrow B$$

between sets, or the intersection and union set

$$A \cap B, \ A \cup B$$

and the power set

$$\mathrm{Pow}(A) = \text{the set of all subsets of } A.$$

Set theory has a disadvantage that will be significant for us later. If we replace the person Jan with the person Erik in the set A, the set

$$A' = \{\text{Erik's dog, Heike's cat}\}$$

is created. If we now form the intersection with the set

$$B = \{\text{Jan's pets}\},$$

then $A \cap B$ and $A' \cap B$ are not equal, although A and A' are isomorphic as sets, because they have only changed into each other by exchanging one element. The intersection is therefore a delicate operation. It is called intensional, as the intersection depends on the respective elements.

The question of the existence of infinite sets played a major role in set theory from the beginning. Dedekind believed that he could prove the existence of an infinite set with the help of the totality of all thoughts possible for a human being. His proof is considered not strictly mathematical and therefore not as correct. Dedekind used infinite sets to construct infinite chains

$$0, \ S(0), \ S(S(0)), \ \ldots$$

of elements in an infinite set with a starting element 0 and a mapping S of the set onto itself, which is called the successor function. Assuming that S sufficiently separates the points of the set, these elements are pairwise distinct from each other and form a model of the natural numbers, usually denoted by $\mathbb{N}$.

Both Dedekind and Cantor considered the collection of all natural numbers in such a model $\mathbb{N}$ as a new actual infinite object of mathematics. For Dedekind, it was clear that there are many different models of the natural numbers. So he proved a theorem stating that all models are isomorphic to each other.[41] This situation is a good example of mathematical objects being different, yet still being identified. Similar uniqueness theorems can be proven for the real numbers $\mathbb{R}$.

After the emergence of Russell's paradox around 1902, which arose from the consideration of the supposed set

$$M = \{x \mid x \notin x\},$$

the naive set theory of Cantor and Dedekind fell into a crisis. From a letter from Hilbert to Frege in 1903, it is clear that Ernst Zermelo had discovered Russell's paradox before Russell did.[42] He also conceived the axiom of choice in 1904. It states that from a family

$$(M_i)_{i \in I}$$

of non-empty sets over an index set I, a tuple of elements m_i can be selected such that $m_i \in M_i$ for all indices $i \in I$. Although this statement seems obvious, it constitutes an independent axiom because the set I can be infinite. Zermelo succeeded in formulating a (conjecturally) contradiction-free axiom system for set theory in 1907. This list was modified by Abraham Fraenkel in 1921 and completed by him in 1930 to the Zermelo–Fraenkel axiom system. It is abbreviated as ZF, or ZFC if the axiom of choice is added.[43]

Set theory became the most popular of the possible foundations of mathematics with this axiom system. This is because sets seem intuitively graspable to many people, even if they are infinite. This perception is deceptive, as the consistency of set theory is not provable by its own means due to Gödel's incompleteness theorems. The set of real numbers $\mathbb{R}$ raises further deep questions. These include the undecidability of the continuum hypothesis. It states that every infinite subset of the real numbers is either bijective to $\mathbb{N}$ or bijective to $\mathbb{R}$. There are models of set theory in which the continuum hypothesis holds and others in which it does not. Only by using additional axioms can it be decided. Thus, the view that infinite sets are well understood objects is at least questionable.

Frege's Concept of Number and Logicism

Frege outlined his ideas on the concept of number in "The foundations of arithmetic" from 1884 and "Basic laws of arithmetic" from 1893.[44] The first book was intended, among other things, to provide a justification for logicism. This idea, which Frege was convinced of in his younger years, asserts that mathematics is not a synthetic a priori science in the sense of Immanuel Kant, but analytic a posteriori. This means that mathematics is based on more fundamental principles and derives its justification from the more fundamental logic. Since arithmetic is based on the principle of complete induction, Frege had to derive this principle from fundamental logical principles and introduce a precise concept of number. For this purpose, Frege used a method called abstraction. In his considerations, he recognised that it is crucial to speak of concepts:

> This suggests to us ... that the number statement contains a statement about a concept.[45]

To obtain number statements, he went beyond logic and used sets to describe concepts. He used the concept of conceptual scope, which goes back to Leibniz. The conceptual scope of F is the set of all objects that fall under the concept F. With these designations, Frege defined the number of a concept F as its conceptual scope. The number 0 can be defined, for example, by a concept that can never be fulfilled and whose conceptual scope is the empty set. As a further illustration, Frege gave the example of the large moons of Jupiter. This concept apparently has the number 4. Since the same number can apply to several concepts, Frege needed Hume's principle:

> The expression "the concept F is equinumerous with the concept G" shall be equivalent to the expression "there is a relation φ, which uniquely assigns the objects falling under the concept F to the objects falling under G".[46]

This principle states that two concepts F and G represent the same number exactly when there is a bijective mapping between the associated scope sets, which we will also denote by F and G for simplicity. An obvious notation for the equinumerosity of the concepts F and G is thus

$$F \cong G.$$

Frege noted in his investigations that this approach does not yet sufficiently define numbers. A famous example of his illustrates the difficulty of distinguishing numbers from objects like the person Julius Caesar. He sought a way out of this situation and had the idea of grouping all representatives of concepts with the same number by abstraction into an equivalence class. In Frege's own words:

> The number which belongs to the concept F is the scope of the concept "equinumerous to the concept F".[47]

The number that belongs to the concept F is thus the set

$$\text{Number to the concept } F = \{G \mid G \cong F\},$$

where a set of concepts G is postulated from a concept F that are equinumerous to F. For example, in today's language, the number 5 is defined as the class of all sets consisting of 5 elements. Although there are many sets with 5 elements, there is only one such class. In the end, Frege had constructed an equivalence class of sets of equal cardinality with his method and thus the problem of defining numbers seemed to be solved.

However, Frege's approach was problematic. He used parts of naive set theory, which he apparently considered as an unproblematic part of logic. Furthermore, he used the axiom of unrestricted comprehension of set theory in a crucial way in the formation of a set of sets in the concept of number. This axiom allows the impredicative formation of sets, i.e., the formation of a set of new objects G from an a priori not determined reservoir, which fulfil a given predicate with a free variable. The use of unrestricted comprehension inevitably led to Russell's paradox, as Frege learned from Bertrand Russell in a famous letter dated June 16th 1902. This letter describes the problem that Russell discovered, but also expresses his appreciation for Frege:

Dear Colleague!

For a year and a half I have known your "Basic laws of arithmetic", but only now have I been able to find the time for the thorough study that I intend to devote to your writings. I find myself in full agreement with you on all main points ... Only in one point have I encountered a difficulty. You claim (p.17) that the function can also form the indefinite element. I used to believe this, but now this view seems doubtful to me, because of the following contradiction: Let w be the predicate, a predicate that cannot be predicated of itself. Can w be predicated of itself? From any answer the opposite follows. Therefore, one must conclude that w is not a predicate. Similarly, there is no class (as a whole) of those classes that do not belong to themselves as a whole ... with you I find the best of what I know from our time, and therefore I have allowed myself to express my deep respect to you.[48]

Russell's paradox can be represented in Frege's thought world by the supposed set

$$M = \{x \mid x \notin x\},$$

which uses the predicate $P(x) = x \notin x$, i.e., x does not contain itself. If then M contains itself, it does not contain itself and if it conversely does not contain itself, then it contains itself. This is obviously a antinomy, i.e., a contradiction.

Frege responded a few days later on June 22nd and indicated that he had recognised the dramatic consequences:

Your discovery of the contradiction has surprised me to the utmost and, almost I would say, dismayed, because thereby the ground on which I thought to build arithmetic is shaken. It seems that the transformation of the generality of an equality into a value-range equality (§9 of my Basic Laws) is not always permitted, that my Law V (§20, p. 36) is false and that my explanations in §31 are not sufficient to secure a meaning for my symbol combinations in all cases ... It is all the more serious as with the disappearance of my Law V not only the basis of my arithmetic, but the only possible basis of arithmetic seems to sink.[49]

Frege was unable to solve the problems caused by Russell's letter and was very unhappy about it. The Russellian antinomy revealed a serious gap in the naive set theory founded by Cantor and Dedekind from 1870 onwards. This was fixed in 1930 through the final establishment of the Zermelo–Fraenkel axiom system and, in particular, by later theories that contain classes that are not sets.

Bertrand Russell tried another way out of the antinomies of set theory, by hierarchising sets into an ascending sequence of infinitely many types, so that the self-reference on which the Russellian antinomy is based, was excluded. Later he published these ideas, which founded type theory, together with Alfred North Whitehead in the three-part book "Principia mathematica" and tried to underpin the idea of logicism.

Frege's logicism was criticised in many ways. The deficits of naive set theory and its use as part of logic are already serious points of criticism. Henri Poincaré was of the opinion that mathematical complete induction is synthetic a priori and therefore cannot be derived from other principles in principle:

This law, which is as inaccessible to analytical proof as it is to experience, gives the actual type of synthetic judgement a priori.[50]

Poincaré's statement can be interpreted to mean that infinite constructs like the natural numbers $\mathbb{N}$ and their arithmetic properties cannot be justified from purely logical principles. Instead, the existence of the infinite set $\mathbb{N}$ and the validity of the proof principle of complete induction must be demanded as an axiom. Since here the consistency of the entire arithmetic is on trial, Poincaré's objections form a serious critique of logicism.

Can logicism be saved? According to the view of a group of neologicists[51] this was already achieved by Frege himself by using Hume's principle, which equates the equinumerosity of sets with their isomorphism. They observed that Frege essentially no longer used conceptual scopes after he had proven Hume's principle. If this principle is postulated as an axiom, then Russell's paradox and Law V can be avoided and Frege's remaining proofs of the properties of the natural numbers are acceptable. In particular, Frege had correctly defined the successor function S and all axioms of Dedekind–Peano arithmetic. However, Hume's principle is not a genuine logical, but a set-theoretical structural principle. This rescue attempt by the neologicists thus does not fully correspond to the original intention of logicism.

Another way of approaching logicism is Russell's type theory. In its current form, it can be understood as the syntax of all mathematics in the form of a higher-level logic. This allows the natural numbers $\mathbb{N}$ to be defined in a way that is closer to Dedekind's definition and to logicism, because the problem of infinity does not occur in the calculus of type theory itself, but only in set-theoretical or categorical models. But even within type theory the question of the consistency of elementary arithmetic remains. Thus, Poincaré's objection remains in some way and logicism must yield to Kant's view that mathematics is an a priori science.[52]

Scientific Languages

2

The search for a universal language of mankind is an old dream. The Babylonian language confusion, which makes it difficult for humanity to communicate, has always posed problems. Over the centuries, there have been different approaches to inventing new artificial interlinguistic planned languages for easier communication in everyday life and formal languages as support for science.

Leibniz pursued such ideas more systematically than many others in his time. In connection with his concept of a universal science, called Scientia generalis, he hoped for formal symbolic calculi that enable correct proofs and calculations and eliminate irrationality, inaccuracies and injustice in all sciences and in human life. This contribution by Leibniz, which went significantly beyond mathematics and logic, was an important step in the history of the philosophical concept of truth. He also coined the motto "theoria cum praxi". It means that science and applications should be closely linked.

Umberto Eco has recorded the history of the most important approaches for interlinguistic planned languages and universal scientific languages in a book.[53] In it, he found that this search was a history of frequent failure, which has produced remarkable partial successes, such as the development of mathematics and logic, computers and artificial intelligence. It is a fascinating and still unsolved task to explore the fundamental limits of this idea.

Leibniz and the Scientia Generalis

Gottfried Wilhelm Leibniz left behind an enormous oeuvre, which has not yet been fully edited to this day. His partly unpublished works only had an appropriate effect at the beginning of the 20th century after the publication of a small part by Louis Couturat.[54]

© The Author(s), under exclusive license
to Springer-Verlag GmbH, DE, part of Springer Nature 2024
S. Müller-Stach, *The Code of Mathematics*, Mathematics Study Resources 11,
https://doi.org/10.1007/978-3-662-69483-1_2

The philosophical system of Leibniz, especially that of his "Monadology",[55] is an alphabet of thoughts reminiscent of Giordano Bruno's concept of monads. His approach consists in reducing thinking to primitive concepts and using these building blocks, as well as with the help of a logical calculus, to reduce the truth of composite statements to simpler questions and ultimately to answer them. Many works by Leibniz can therefore—in addition to his works on theology and metaphysics—be seen as the beginning of epistemology with the help of symbolic methods. He believed that human thinking—in contrast to his opinion of God's intuitive and comprehensive perception—due to the limits of human reason, relied on symbolic knowledge. With this mutual positioning of theology and mathematics, Leibniz was in the tradition[56] of Christian theology.

Leibniz anticipated the ideas of logicians like Gottlob Frege in his thinking and built on insights that were in circulation long before his time. Although algorithms have been used since antiquity, such as the Euclidean algorithm in antiquity or the calculations of Easter in the Middle Ages, it is only the Majorcan scholar Ramón Llull (Latin Raimundus Lullus) in the 13th century who is attributed with the concept of deductive and algorithmic thinking, which is the basis of proofs and calculating machines. Leibniz, in addition to his numerous other interests, not only dealt intensively with logic, but also with other parts of mathematics. In many ways, Leibniz's dream of the Scientia generalis was also a dream of a generalised mathematics.

Leibniz refers in his "Dissertatio de arte combinatoria"[57] from 1666 explicitly to the book "Ars magna"[58] by Llull from 1290. Leibniz was also familiar with the ideas of René Descartes, Thomas Hobbes and John Wilkins.

Hobbes had recognised in his work "De corpore"[59] that thinking in a generalised sense can be equated with calculating. Descartes had already dreamt of a universal philosophical language before Leibniz, as he wrote in a letter to the number theorist and monk Marin Mersenne on November 20th 1629:

> One should arrange all thoughts methodically, just as the natural sequence of numbers is methodically arranged. Just as one can learn in a single day in any foreign language to name and write all numbers to infinity, the numbers, which certainly form an endless series of combinations, one must find the possibility to construct all words that are necessary to express everything that comes to the human mind and can come ... The invention of such a language depends on true philosophy.[60]

John Wilkins published a book[61] in 1668 about a universal philosophical language, which should be superior to natural language. Most of the other universal languages conceived at the time were primarily designed for recording knowledge and communication and anticipated today's interlinguistic planned languages.[62]

Leibniz thought more deeply than many others about a comprehensive universal science, which he called Scientia generalis. It should have an underlying universal scientific language, which we will call Lingua universalis in this text. This was for Leibniz a tool to carry out thinking in a symbolic way and to perform conclusions and proofs with a computational method by term substitutions within the calculus. This art of judgement was called Ars judicandi at that time. The proximity of proof and calculation, which will become important for us, was thus already present in the ideas of Leibniz, Descartes, Hobbes and Wilkins. Furthermore, it should be possible to develop new creative thoughts and concepts in the form of an art of invention (Ars inveniendi) by using this language.[63]

Presumably due to time constraints, Leibniz never fully developed these ideas himself in a satisfactory way. However, he regularly described them in his correspondence, for example in a letter to Nicolas Rémond dated January 10th 1714:

> I dare to add something, namely, if I had been less distracted or if I were still younger or had the help of young, talented people, then I would have the hope of bringing out a kind of "general correctness theory", in which all truths of reason would be reduced to a kind of calculation. This could simultaneously be a kind of universal language or universal script, but infinitely different from all those that have been proposed to date; for the signs and the words themselves would guide reason here, and the errors (with the exception of those about a fact) would only be calculation errors. It would be very difficult to form or invent this language or characteristic, but very easy to learn it without any dictionaries. It also served to estimate the degrees of probabilities (if we do not have enough data to arrive at certain truths) and to see what is needed to supplement here.[64]

The calculus dreamed of by Leibniz was realised within mathematical logic in the 19th and 20th centuries by Frege and others. However, in its intended generality, it has not been constructed to this day and was not applied by Leibniz himself within his metaphysics. One reason for this may be that his time would probably not have accepted a calculus of clearly derivable truth anyway.

Leibniz and Mathematics

Although Leibniz was a self-taught mathematician, he had significant ideas. For many years he commissioned the construction of a calculating machine with a stepped drum for the four basic arithmetic operations and it can be speculated whether he dreamed of a powerful universal machine to put his symbolic calculus into practice and thus answer all philosophical questions rationally.

Leibniz introduced computing based on the binary number system with the digits 0 and 1, which would become the basis of digital computers. Equally successful was his invention of the differentials dx and the differential quotients

$$\frac{df}{dx}$$

in the form of a calculus. The differential dx symbolises an infinitesimally small quantity. The calculus[65] of Leibniz has been maintained to this day and included the following calculation rules:

$$d(f + g) = df + dg \qquad \text{(additivity)}$$
$$d(af) = adf \qquad \text{(linearity)}$$
$$d(fg) = fdg + gdf \qquad \text{(product rule)}$$

$$d\left(\frac{f}{g}\right) = \frac{gdf - fdg}{g^2} \qquad \text{(quotient rule)}$$

$$d(f \circ g) = \frac{df}{dg}dg \qquad \text{(chain rule)}.$$

In the development of infinitesimal calculus, Leibniz was in a sometimes bitter competition with Isaac Newton and his theory of fluxions. Although he was inferior to Newton and the Royal Society on some occasions, his status in science today is on par with Newton's.

Calculus presupposes a concept of real numbers. Such a concept was certainly held by Leibniz and others in his time. However, it was not until the 19th century that Cantor and Dedekind precisely defined the set $\mathbb{R}$ of real numbers and proved theorems about it. Modern calculus, with its precise concept of limits, was only established in the 19th century by Karl Weierstraß and others. Only much later could a mathematically flawless definition of infinitesimally small quantities be given with the non-standard analysis of Abraham Robinson, which revived the calculus of Leibniz.

Leibniz formulated some remarkable statements about real numbers. He already knew the difference between rational, irrational, algebraic and transcendental numbers and proved results about the transcendence of certain integrals of differential forms.[66]

The Rise of Mathematical Logic

Aristotle is considered the founder of mathematical logic through his study of syllogisms. These consist of a chain of statements, also called a mode (Latin modus):

Major premise: All B are C
Minor premise: All A are B
Conclusion: All A are C.

In this example, the modus barbara is given. Aristotelian logic included a whole range of other such modes and had a dominant position for centuries in the scholastic tradition until the Middle Ages. Aristotle already knew the law of excluded middle, i.e., the principle underlying proof by contradiction.[67]

From today's perspective, the Aristotelian modes are only fragments of modern logic. Aristotelian logic was revived in the 20th century by Jan Łukasiewicz and compared with then-modern theories. After ancient culture, logic was practised as a fundamental cultural technique in Europe, Asia and the Islamic world. These discoveries anticipated many modern developments, but often remained isolated. In early modern Europe, it was primarily Leibniz who had the most significant insights in logic. He provided tools of mathematical logic in the form of a calculus, which anticipated the algebraic logic of the 19th century, including a version of quantifiers and thus predicate logic. Leibniz anticipated modal logic in its beginnings, which is related to his theory of possible worlds. However, all this remained hidden to most people of his time because Leibniz did not publish his thoughts on logic.[68]

Mathematical logic experienced after Leibniz a great upswing in the 19th century, particularly at universities in Europe and the United States. The two English logicians George Boole and Augustus de Morgan, as well as somewhat later Gottlob Frege and Giuseppe Peano, particularly shaped the basic concepts of algebraic propositional logic and the beginnings of predicate logic.[69] Logic as a mathematical theory only fully unfolded in the 20th century.

Frege recognised from examples that natural language is not a good basis for the foundations of mathematics and therefore developed mathematical logic as

The science of the most general laws of truth.[70]

He shaped mathematical logic in his "Begriffsschrift".[71] Frege's variant of a Leibnizian language included logical symbols such as the judgement stroke

$$\vdash A$$

as well as the universal quantifier and the existential quantifier

$$\forall x, \exists x.$$

With this, he introduced the first deductive system with a formal language in the history of logic, which we would today call second-order predicate logic or simply second-order logic.[72] Some of the common logical notations today, however, are more due to Peano than to Frege.

Leibniz had a remarkable influence on Gottlob Frege, Giuseppe Peano and Kurt Gödel. At the beginning of the "Begriffsschrift", Frege explicitly referred to Leibniz as a role model:

Leibniz too recognised the advantages of an appropriate notation, perhaps overestimated. His idea of a general characteristic, a calculus philosophicus or calculus ratiocinator was too gigantic for the attempt to realise it to get beyond mere preparations.[73]

The beginning of the classical logic of the 19th century was the calculus of propositional logic, to mathematically handle logical statements. In it, there are the possible truth values

$$\top \text{ (true)}, \bot \text{ (false)}$$

and statements are connected to new statements through the connections

$$A \wedge B \text{ (and)}, A \vee B \text{ (or)},$$

$$\neg A \text{ (negation)}, A \Rightarrow B \text{ (conditional or implication)}.$$

A	B	$A \Rightarrow B$	$\neg A \vee B$
true	true	true	true
true	false	false	false
false	true	true	true
false	false	true	true

Truth table.

While the conjunction $\wedge$ is familiar to us in everyday life, the disjunction $\vee$ is often used in an exclusive sense. In mathematical logic, however, the statement $A \vee B$ means that at least one of the two statements A or B is fulfilled, i.e., they can both be fulfilled at the same time. In classical propositional logic, calculation rules such as

$$\neg A \vee \neg B = \neg(A \wedge B) \text{ and } A \Rightarrow B = \neg A \vee B$$

apply, so that some symbols are in principle redundant. Proofs for such calculation rules can be conducted with truth tables.

The calculation rules of propositional logic for the unary or binary operations $\wedge, \vee, \Rightarrow$ and $\neg$ result in an algebraic structure, which in honour of George Boole is also called Boolean algebra. Such a structure Ω is in general not a ring in the usual mathematical sense, even though the operations $\vee$ and $\wedge$ have similar properties to addition and multiplication. In fact, $\vee$ and $\wedge$ allow a lattice structure[74] on Ω by the union or intersection of subobjects. The objects $A \Rightarrow B$ and $\neg A$ are called exponential object B^A and negation of A. On a Boolean algebra there also exists a partial order $\leq$, which can be defined by the relation

$$A \leq B \text{ exactly when } A = A \wedge B$$

through the operation $\wedge$. In Ω the equation

$$\neg A \vee B = (A \Rightarrow B)$$

holds and $\neg$ is an involution:

$$\neg\neg A = A.$$

Heyting algebras are generalisations of Boolean algebras, in which these two rules are abandoned. In them only[75]

$$(\neg A \vee B) \leq (A \Rightarrow B) \text{ and } A \leq \neg\neg A$$

apply. Heyting algebras play a major role in intuitionism.

The Question of Truth

Aristotle also dealt with the concept of truth. In his "Metaphysics"[76] he made fundamental considerations on which two significant concepts of truth are based. The first of these is the adequation theory of truth. It sees the concept of truth in the agreement of thinking with reality. Thomas Aquinas expressed this as follows:

Veritas consistit in adaequatione intellectus et rei.[77]

By thinking, we mean the contents of consciousness and thoughts in our brain. We regard reality as the material physical reality together with its natural laws, but also as other phenomena, which may not be reducible to physics, such as questions of existence and metaphysics.

The correspondence theory of truth—related to the adequation theory of truth—replaces thoughts with the corresponding linguistic propositions in our thinking, i.e., the formulations of our thoughts. Aristotle gave an explicit example of the correspondence theory:

Not because our opinion, that you are white, is true, are you white, but because you are white, we tell the truth, by asserting this.[78]

The correspondence theory in its formulation thus means the correspondence between statements and facts of reality. Both theories are to be questioned, because they connect two a priori incomparable categories such as firstly thinking and the contents of consciousness in our brain—or associated propositions—and secondly reality in an inexplicable way. Problems lie in the relationship of the truth bearer, which could either be a thought or a linguistic statement, with the truth maker, i.e., a fact of reality, because it is not obvious how facts could be determined without already having a definition of truth. For these reasons, there have always been controversial discussions about the correspondence theory. Thomas Hobbes wrote in his "Leviathan" around 1651:

For true and false are attributes of speech, not of things.[79]

When such concepts of truth are applied to mathematics, serious problems arise, because mathematical concepts usually have no correspondence in reality. Moreover, the concept of reality itself is problematic, because it is not clear what exactly is meant by it. When people talk about reality, they are usually guided by aspects of matter and there is a basic trust in the concreteness of this physical reality. Upon deeper reflection, however, we encounter doubts and will come to the conclusion that even the physical reality has a very abstract nature, which we cannot comprehend as a whole and therefore can only describe with mathematical language. An example are electromagnetic fields, which spread in complete vacuum according to mathematical laws. Such fields are only tangible through measurements or their effect on other physical objects. Phenomena of quantum mechanics such as the dualism between particles and waves as well as the Heisenberg uncertainty relation elude even more our intuition. The abstractness of reality and its characteristics is the subject of the philosophical question of realism and allows for contradictory positions.[80]

These problems with the adequation and the correspondence theory of truth had various consequences. On the one hand, this has led to the search for further theories of truth that are free from connections with any kind of reality. An example of this is the coherence theory of truth. It states that for the truth of statements of a new theory, the coherence within the scope of the already existing or underlying theoretical structure is primarily important. A conditio sine qua non for this is again the consistency—i.e., the contradiction-freeness—of the theory under consideration itself. The coherence theory is also controversial in philosophy due to the existence of ambiguous coherent systems, as Bertrand Russell already noted.[81] We will return to whether this view is at least fruitful within mathematics. We do not provide an exhaustive description of all existing theories of truth.

Leibniz and Truth

Leibniz had his own conception of the correspondence theory. The concept of truth for him was a property of thoughts, not things. In the "Dialogus", he accordingly wrote in 1677:

> Veritatem esse cogitationum non rerum.[82]

The Leibnizian theory of truth uses the distinction between factual truths and rational truths. The truth of propositions is examined by decomposing composite statements (notio composita) into simpler, indivisible statements using the symbolic calculus of the Lingua universalis. Leibniz wrote about this in his "Monadology":

> There are also two kinds of truths: those of reasoning, which are necessary and whose opposite is impossible, and those of fact, which are contingent and whose opposite is possible. If a truth is necessary, one can find the reason for it by analysis, by resolving it into simpler ideas and truths, until one reaches the initial ones.[83]

Factual truths apply in our world and are gained through experience. Rational truths apply in all possible worlds. Leibniz had invented a theory of possible worlds to locate other realisations of truth. It served him to represent the reality in which we live as the one preferred by God as the best of all possible worlds.

Leibniz incidentally spoke of God as a being that does not need the symbolic language because it can directly grasp the truth. Moreover, in his opinion, God could experience the world and its history, i.e., the concept of space-time, as a whole all at once. If God is not considered as a being, but is understood as a universal principle, synonymous with the world, then Leibniz's perspective leads to the realisation that, for us ordinary people, reality could only be a semantic internal representation, because we do not fully comprehend it. While Leibniz's image of God naturally eludes all proof, his theory of possible worlds was in a way realised in the 20th century through modal logic.

Truth and Kantian Judgements

In his considerations on the concept of truth, Immanuel Kant significantly developed Leibniz's theory of truth, without explicitly mentioning Leibniz.[84] Kant used the term judgement for statements whose truth is to be investigated. He made an epistemic distinction between a priori judgements—necessary and universally valid judgements independent of experience—and a posteriori judgements—experiential judgements through empirical knowledge—, which in a way goes back to the Aristotelian dualism between proteron (condition) and hysteron (conditioned).

Mathematical axioms are good examples of a priori judgements. A related example underlies any attempt at a proof of God, because the existence of divine beings cannot be derived from experience or observation of the world, even if some people claim this. Serious proofs of God, like the proofs of Anselm of Canterbury and Kurt Gödel, are based on logical arguments and use an axiom that connects divine beings with maximal positive properties.

Immanuel Kant further distinguished between analytical and synthetic judgements,[85] which express a reason for the validity of truth. An analytical judgement is an explanatory judgement, i.e., it can be directly inferred from the definition of the object or concept under consideration. In contrast, a synthetic judgement is an extension judgement, i.e., it can only be explained by means of a further justification. An example of an analytical judgement is:

It is dark at night.

This is because it is precisely the definition of night that it is dark at this time, unless there is a light source present.

The pairs of properties analytical-synthetic and a priori-a posteriori are initially in principle independent of each other. In this system of thought, Leibniz's rational judgements become analytical judgements a priori and the factual judgements become synthetic judgements a posteriori. Analytical judgements a posteriori are excluded, as analytical judgements are always a priori,[86] so that a third category of synthetic judgements a priori remains, which according to Kant is an existing category.

Kant attributed the property pairing of synthetic a priori to mathematics. Mathematics is certainly a priori, as its judgements are not empirical knowledge. Kant claimed that mathematics is synthetic, as it cannot be derived from anything else. As an example, he used the equation

$$7 + 5 = 12.$$

It is the question whether all possible such equations, i.e., the entire arithmetic, directly follow from the definition of numbers or whether they can be derived from other principles. Such a principle could be a logical principle or a mathematical axiom. Leibniz would have spoken of a sufficient reason in this context.

From the question of whether mathematics is synthetic or not, logicism emerged, which aimed to reduce mathematics completely to logic, so that it would consequently be analytic. Especially Gottlob Frege and Bertrand Russell propagated logicism and thus rejected the view of Kant. Willard van Orman Quine and others later had doubts that the distinction between analytic and synthetic judgements is clear-cut and concluded that logicism is not sufficiently justified.[87]

In mathematics, the Kantian distinction between analytic and synthetic judgements is not common. But it is quite relevant, as synthetic judgements encounter us through the postulation of new axioms and the introduction of primitive concepts that cannot be derived from the remaining assumptions.

Truth, Sense and Meaning According to Frege

We have already touched on the fact that Frege dealt with the concept of equality and in this context examined the concepts of sense and meaning of linguistic statements. He also thought about how the truth of statements can be proven and how the statements and their potential truth are related to the thoughts and the content of our consciousness. The concepts truth, sense and meaning all belong to the field of semantics, which is located in the language philosophy and linguistics ultimately going back to Frege. With his investigations, in which he carefully distinguished all occurring concepts from each other, he became a significant philosopher of modern times after his works on logic.[88] Frege was of the opinion that the equation

$$A = A$$

is an a priori true analytic statement (in the sense of Kant). However, the statement

$$A = B$$

is an extension of a completely different form, possibly synthetic and not automatically a priori. Frege explained the difference between the concepts of sense and meaning using this example, as we have already explained using the triangle example.

Frege particularly intensively dealt with the related concept of truth. As early as 1892, he formulated the following thought:

> One could even say: "The thought that 5 is a prime number is true." But if one looks more closely, one notices that actually no more is said than in the simple sentence "5 is a prime number".[89]

Because of such statements, Frege is often accused of a redundancy-theoretical attitude, which considers truth irrelevant for the meaning of statements. Basically, he considered truth undefinable at that time, as he noted in a note from 1906:

What is true, I consider unexplainable.[90]

However, he continued his reflections and tried to precisely explore aspects of truth. He wrote in an essay in 1918:

So with every property of a thing, a property of a thought is connected, namely the truth.[91]

The bearer of truth lies, according to Frege's opinion, in the statements of thoughts and not in the understanding of man or in reality. The moment the truth of a thought is asserted or shown, man makes a judgement. Frege wrote about thinking and the world of thoughts:

Thoughts are neither things of the outside world nor ideas. A third realm must be recognised.[92]

Frege's realm of thoughts as a separate place of objective facts reminds us of the Platonic world of ideas. At an earlier point, Frege had written on this topic:

But if the subjective has no place, how is it possible that the objective number 4 is nowhere? Now I assert that there is no contradiction in this. It is indeed the same for everyone who deals with it; but this has nothing to do with spatiality. Not every objective object has a place.[93]

So was Frege a Platonist or not? Similarly, one can ask why Frege mentioned Kant, Leibniz, Mill and others in his book "Foundations of arithmetic" and elsewhere, but not German idealism, represented by Fichte, Schelling and Hegel, who had advanced conceptual philosophical thinking in their works.

However, Frege was very sceptical about concepts like existence, as his "Dialogue with Pünjer on existence" proves. Thus, he was certainly rather sceptical towards both Platonism and idealism. He is now considered the founder of analytical philosophy and had an influential student in Rudolf Carnap, who was a member of the Vienna Circle led by Moritz Schlick.[94] The members advocated logical empiricism and vehemently rejected metaphysics.

We will get to know the ideas of Hermann Weyl, who pondered the symbolic construction of reality in his philosophical works and had no fear of contact with idealism and metaphysics. He maintained a correspondence with Edmund Husserl and was an admirer of Johann Gottlieb Fichte. In today's time, analytical philosophy and metaphysics are not easily distinguishable and their differences lie more in the approach than in the philosophical topics. Already Leibniz had recognised that scientific metaphysics necessarily must be based on a deductive calculus.

Tarski's Theory of Truth

When investigating the truth of statements in non-scientific contexts, there are fundamental difficulties associated with the concept of truth. This is due to self-referential
sentences in natural languages such as the paradox of Epimenides

A Cretan says: All Cretans lie[95]

or sentences like the paradox attributed to Bertrand Russell

The barber who shaves all men who do not shave themselves.

These well-known paradoxes have been widely questioned[96] and it is better to use a
self-referential statement of the form

A: The statement A is false

in which the paradox comes into sharper focus. In formal languages, such problems
can be avoided, as Alfred Tarski has shown in his language-analytical theory of truth.

His theory can be applied to formal object languages such as those underlying
mathematics. For this purpose, an interpretation of the statements in a metalanguage
M is used, which usually contains L as a sublanguage, extending beyond a given
formal language L. By separating the two languages L and M, self-reference is
avoided and a concept of truth in the form of a truth predicate $T(x)$ in M is defined.

Tarski has given a syntactic adequacy condition to be fulfilled in biconditional
form that reminds of Aristotle:

The sentence "Snow is white" is true if and only if snow is white.[97]

Here, the left part of the sentence within the quotation marks is an atomic statement
p of the formal object language L. On the left side, the truth predicate T of the
metalanguage M applied to p and is equivalent to the right-hand side, which consists
of the translation $\tilde{p}$ of p into the metalanguage M. So, a bit more formally, in M it
holds

$T(p)$ exactly when $\tilde{p}$,

where a standard name for p must be substituted as the argument of $T(x)$.[98]

The adequacy condition is not sufficient as a definition of truth, but only a necessary prerequisite. The actual definition of truth is made via the satisfiability in a
suitable model, which is also called semantics. In it, the metalanguage M is used
as a deductive system. The most natural example of mathematical semantics is the
formal object language L_{ar} of Dedekind–Peano arithmetic and its interpretation in
the set-theoretical standard model $\mathbb{N}$ or the various non-standard models. In these

models, the metalanguage M is usually given by the Zermelo–Fraenkel set theory with first-order predicate logic. From this it follows that the concept of truth corresponds to the concept of provability in a deductive system corresponding to the model which has a richer formal metalanguage M than L.

Already Leibniz and Frege recognised that methods of mathematical logic can be used in inferences to prove the truth. Later, an important reason was found why the proof must be carried out in a strictly richer metalanguage. In Dedekind–Peano arithmetic, self-referential constructs of the form

A: The statement A is false

can be cleverly generated with the trick of Gödel numbering. This results in the (first) incompleteness theorem by Kurt Gödel and the related theorem on the undefinability of truth by Alfred Tarski. These theorems demonstrate the existence of undecidable statements in formal object languages L, which in L are neither provable nor refutable, as well as the undefinability of the truth predicate in L itself.[99]

What Became of the Lingua Universalis?

The development of logic and formal scientific languages has brought us within mathematically oriented sciences close to a realisation of a Leibnizian Lingua universalis. In contrast, outside of mathematics, the attitude is widespread that this idea is idealistic, cannot encompass all human thinking and many truths in the sciences and in everyday life fundamentally cannot be captured with a formal language. Even the optimistic Leibniz believed in the limits of knowledge and thus of the Lingua universalis. Through his mill example, he expressed that phenomena such as the consciousness of humans and animals and other qualia are difficult to explain. In the life sciences and in philosophy, the phenomenon of consciousness is controversially discussed, so that the exact boundaries in this case remain unclear to this day.

Digitisation has brought forth new algorithms in recent years that can excellently imitate human thinking. In almost all areas, the formal verifiability of knowledge is necessary due to large amounts of data and made possible by intelligent digital assistant systems. However, most of the systems that have emerged do not yet have the judgement that Leibniz had envisaged. It is an interesting and difficult question to explore the nature of universal scientific languages. They certainly extend—despite the mentioned counter-positions—far into areas beyond mathematics and computer science. It remains an exciting question to explore their limits and the connection with the foundations of our world.

Mathematical Thinking

3

Statements like "I was always bad at maths" or "What is there still to research in mathematics?" are commonplace in the media and at private meetings, regularly causing unpleasant feelings for professionals. Many people are proud of their lack of knowledge in mathematics, even if they are ashamed of their other intellectual deficits. Even if we cannot expect mathematics to be important to all people, we should continue to develop the mathematical curriculum in schools and try to change this situation.

This is not hopeless, as the mathematical way of thinking is more similar to that in art. Godfrey Harold Hardy, a famous British number theorist, wrote perhaps the best-known book about the inner view of mathematics. It is titled "A mathematician's apology".[100] In this book, the beauty of mathematics and its kinship with art play a special role. Hardy emphasises therein that the usefulness of mathematics in applications is no measure of its sense and quality. He points out that areas like number theory possess an inner beauty and have no applications. On this point, Hardy was mistaken, as number theory now has many applications, such as blockchains and public-key cryptography.

Outside of number theory, the way of thinking of mathematics can also be impressively explained using the world of transfinite sets, the universal, free constructions of algebra and the description of topological space forms.

Natural Numbers

Number theory, also called arithmetic, is a field of mathematics that deals with various types of numbers and their properties. This includes in particular the natural numbers, the integers and the algebraic numbers, which are the roots of polynomials with integer coefficients. Number theory and geometry existed already in antiquity. Euclid and Diophantus were leaders in these fields.

© The Author(s), under exclusive license
to Springer-Verlag GmbH, DE, part of Springer Nature 2024
S. Müller-Stach, *The Code of Mathematics*, Mathematics Study Resources 11,
https://doi.org/10.1007/978-3-662-69483-1_3

The multi-part work "Arithmetica" by Diophantus[101] contains a fascinating arithmetic problem which consists in the search for integer solutions of diophantine polynomial equations. The most famous equation of this kind is Fermat's equation

$$x^n + y^n = z^n.$$

According to a theorem by Andrew Wiles from 1993, it only has strictly positive integer solutions when $n = 1$ or 2, as Pierre de Fermat had suspected. Solutions for the case $n = 2$ can already be found on Babylonian clay tablets[102] around 1800 BC, such as the famous Plimpton 322 tablet. Slightly more general than solutions of a polynomial equation are algebraic varieties, which arise as simultaneous zero sets of several polynomials. These frequently occur in mathematical modelling.[103]

Euclid is the originator of the Euclidean algorithm, the prototype of recursive thinking. It is based on the fact that two natural numbers a and b (with $b \neq 0$) can be written in the form

$$a = qb + r \text{ with } 0 \leq r < b,$$

where r is the remainder of a modulo b and qb is the maximum multiple of b that is still less than or equal to a. The Euclidean algorithm replaces the pair (a, b) with the pair (b, r) in each step and stops as soon as the remainder r becomes zero for the first time. The number b in the last pair $(b, 0)$ is the sought-after greatest common divisor of a and b. The algorithm also works if b is larger than a, because then $r = a$ and $q = 0$ and the pair (a, b) is replaced by (b, a) in the first step. Since $r < b$ in each step, the process terminates after a finite number of iterations. This algorithm can be easily programmed and executed quickly in any common programming language.[104]

The calculation of q as the floor of $\frac{a}{b}$ can be omitted, because the remainder r is reached when b is subtracted from a until a positive number less than b is obtained. This variant is called the fast Euclidean algorithm.

We are interested in the properties of the entirety $\mathbb{N}$ of all natural numbers with 0 as the smallest number among them. The infinite sequence

$$0, 1, 2, 3, 4, 5, \ldots$$

is based on the principle of counting on. It is generated by the successor function S, which maps each number n to the following number

$$S(n) = n + 1.$$

It holds

$$1 = S(0)$$
$$2 = S(S(0))$$
$$\vdots$$

If we write

$$S(n) = n + 1,$$

this suggests an addition mapping. The natural numbers indeed possess an addition

$$m + n$$

and a multiplication

$$m \cdot n.$$

These two arithmetic operations can be precisely defined using the successor function S and recursion.

Very large numbers can be easily written in mathematics, for example by using powers of ten

$$10^2 = 100$$
$$10^3 = 1000$$
$$10^6 = 1000000, \text{ one million}$$
$$\vdots$$
$$10^{100} = 1\underbrace{000\ldots000}_{100 \text{ zeros}}, \text{ a googol}$$
$$\vdots$$

The infinite sequence of natural numbers has no counterpart in the real world. This can be easily understood, because the number of particles in the universe is very large, but finite. According to reasonably realistic estimates based on current models, this number in the visible universe is about the order of magnitude 10^N with N between 80 and 90.

Prime Numbers

The sequence of ever larger numbers is not mysterious in itself. Only when we consider a concept like divisibility does the world of prime numbers unfold from the natural numbers

$$2, 3, 5, 7, 11, 13, \ldots, 101, 103, \ldots$$

These are exactly the natural numbers that are divisible only by 1 and themselves. There are infinitely many prime numbers and every natural number can be uniquely written as a product of such. A well-known proof for the infinity of the set of all prime numbers was already given by Euclid.[105] Another wonderful proof considers the Fermat numbers of the form

$$F_m = 2^{2^m} + 1$$

for $m = 0, 1, 2, \ldots$. The first of these numbers are

$$F_0 = 3$$
$$F_1 = 5$$
$$F_2 = 17$$
$$F_3 = 257$$
$$F_4 = 65537.$$

Fermat had speculated in a letter to the mathematician Bernard Frenicle de Bessy in 1640 that all Fermat numbers are prime numbers. However, the number $F_5 = 641 \cdot 6700417$ is not a prime number. We claim, nevertheless, that all these numbers are coprime to each other, i.e., they have no common divisor. This implies the infinity of prime numbers, because the prime factors in the numbers F_n are all different from each other. For this claim, we first show the formula[106]

$$F_m = F_0 F_1 \cdots F_{m-1} + 2.$$

If F_k and F_m with $k < m$ had a common divisor p, then p would also divide the 2. This is a contradiction, because all F_m and thus also p are odd.

Computers today can handle numbers of the order of several hundred decimal places and with great effort investigate the factorisation of such numbers into prime factors. The problem of finding the factorisation of a number is presumably much harder than checking whether a given number is prime, because there are prime number tests that have a limited runtime. An example of a prime number with 50 digits is the number

$$p = 53542885039615245271174355315623704334284773568199.$$

The currently largest known prime numbers are the rarely occurring Mersenne primes of the form $2^p - 1$ with p prime. At the time of writing this book, the number

$$2^{82.589.933} - 1$$

with 24.862.048 digits was the largest proven Mersenne prime number.[107]

Since the time of Bernhard Riemann, a surprising amount is known about the distribution of prime numbers. Although he has a comparatively small oeuvre, his works were all the more influential. Among them is the formulation of the Riemann hypothesis[108] in an unpublished work of less than seven pages. It states that the zeros of the complex-valued Riemann ζ-function

$$\zeta(s) = \sum_{n=1}^{\infty} n^{-s}$$

are either the so-called trivial zeros at the negative even numbers $n = -2, -4, -6, \ldots$ or lie on the vertical line with the equation $\text{Re}(s) = \frac{1}{2}$ in the complex plane. This hypothesis has not yet been proven, although there is plenty of evidence for it. The assumption of further zeros on $\text{Re}(s) = \frac{1}{2}$ was a great insight by Riemann. In the aforementioned short work, he proved a connection between the location of the zeros of $\zeta(s)$ and the distribution of prime numbers by applying the methods of Fourier theory to the ζ-function. This resulted in an analytical formula for the number $\pi(x)$ of prime numbers below a limit x, which can be expressed with the help of the non-trivial zeros. Only in 1932 was a formula discovered in Riemann's posthumous notes with which he had calculated zeros of the ζ-function.[109]

While such formulas are very precise and can in principle find every prime number, there are also results like the prime number theorem, which predicts the asymptotic distribution of prime numbers. It states that $\pi(x)$ is well approximated by the function $\frac{x}{\log(x)}$ and—even somewhat better—by the logarithmic integral $\mathrm{Li}(x) = \int_2^x \frac{dt}{\log(t)}$. This theorem was proven in 1896 by Salomon Hadamard and Charles–Jean de La Vallée Poussin. Assuming the Riemann hypothesis, it can be shown that the error is bounded by

$$|\pi(x) - \mathrm{Li}(x)| < \frac{\sqrt{x}\log(x)}{8\pi}.$$

The zeros and special values of the Riemann ζ-function and its generalisations in the form of L-functions carry information about numerous arithmetic objects.[110]

The Structure of the Number System

From the natural numbers $\mathbb{N}$, the entire number system from school can be built. This includes the transition from $\mathbb{N}$ to the integers $\mathbb{Z}$ and from there to the rational numbers $\mathbb{Q}$, the fractions. Finally, we want to construct the real numbers $\mathbb{R}$ from $\mathbb{Q}$. The transition from $\mathbb{N}$ to $\mathbb{Z}$ is called the Grothendieck group. For this, we imagine any integer, especially the negative numbers, as a difference of natural numbers

$$a - b, \quad \text{with } a, b \in \mathbb{N}.$$

This representation is not unique, so we equate pairs (a, b) with the same difference, such as

$$(1, 2) = (2, 3).$$

Negative numbers $-b$ correspond in this view to the pairs $(0, b)$. In general, we identify

$$(a, b) = (c, d) \text{ exactly when } a + d = b + c.$$

The equation on the right side only contains natural numbers due to this rearrangement. Consequently, the integers $\mathbb{Z}$ are pairs of natural numbers that are identified in a precise manner. They form a commutative ring, i.e., they carry an addition

$$(a, b) + (c, d) = (a + c, b + d)$$

and a multiplication

$$(a, b) \cdot (c, d) = (ac + bd, ad + bc),$$

both of which are connected by the distributive law.

When transitioning from $\mathbb{Z}$ to $\mathbb{Q}$, fractions

$$\frac{a}{b}$$

are formed from integers in a similar way, where a and b are integers and $b \neq 0$. Here too, we work more precisely with pairs (a, b) with $b \neq 0$ and identify

$$\frac{a}{b} = \frac{c}{d}, \text{ if } ad = bc.$$

The usual rules of fraction calculation apply:

$$\frac{a}{b} + \frac{c}{d} = \frac{ad + bc}{bd}$$
$$\frac{a}{b} \cdot \frac{c}{d} = \frac{ac}{bd}.$$

This construction presented here can be carried out for every commutative ring R free of zero divisors instead of $\mathbb{Z}$ and is called the field of quotients of R. Like the Grothendieck group, it first appeared in an unpublished manuscript[111] by Richard Dedekind. Of course, these calculation rules for the transition from $\mathbb{Z}$ to $\mathbb{Q}$ were in use long before Dedekind and have long been part of the school curriculum.

Fields of quotients are special cases of a more general construction, in which all objects in a multiplicatively closed subset S of a ring R are made invertible. This is called the localisation of a ring after S, denoted $S^{-1}R$. In the field of quotients of a commutative ring free of zero divisors, localisation is done after all ring elements that are not equal to 0. The Grothendieck group and more generally the localisations are examples of quotient constructions, where elements are identified so that new mathematical objects can be created.

From the rational numbers $\mathbb{Q}$, the real numbers $\mathbb{R}$ can be constructed using Dedekind cuts, which are certain subsets of $\mathbb{Q}$, or with Cauchy sequences. The approach with Cauchy sequences is called completion and is defined as a quotient after the null sequences with respect to the archimedean norm $| - |$, i.e., the absolute value. This method leads to the more exotic p-adic numbers $\mathbb{Q}_p$, which are a completion of the rational numbers $\mathbb{Q}$ with respect to a non-archimedean p-adic norm $| - |_p$ for a prime number p. The p-adic norm of p is given by $1/p$, so that high powers of p, unlike in the archimedean case, have a small norm.[112]

Transfinite Numbers

Georg Cantor defined a fascinating world of infinite numbers using set theory, which are called ordinal and cardinal numbers. Ordinal numbers are well-ordered sets, i.e., totally ordered[113] sets, such that every non-empty subset has a smallest element. Cardinal numbers are certain ordinal numbers, the smallest among all ordinal numbers that possess the same cardinality. We want to explain this using examples.

All natural numbers are simultaneously ordinal numbers and cardinal numbers. John von Neumann represented them as sets of the respective preceding numbers:

$$0 = \emptyset, 1 = \{0\}, 2 = \{0, 1\}, \ldots, n + 1 = \{0, 1, \ldots, n\}, \ldots$$

The smallest infinite ordinal number is ω. It corresponds to the normal ordering

$$0 < 1 < 2 < 3 < \cdots$$

of the natural numbers. The associated cardinal number of $\mathbb{N}$ is denoted by $\aleph_0$. One of the most remarkable insights of Cantor was that counting can continue beyond the ordinal number ω:

$$\omega < \omega + 1 < \omega + 2 < \cdots < 2\omega < \cdots < \omega^2 < \cdots < \omega^\omega < \cdots < \varepsilon_0 < \cdots$$

In addition, Cantor introduced an addition and multiplication of ordinal numbers, both of which are generally not commutative. The addition $\alpha + \beta$ is defined by "placing α and β side by side". For example,

$$\omega + 1 = \omega \cup \{*\} \neq \omega,$$

where $*$ is greater than all $n \in \omega$. On the other hand, it is apparently true that

$$1 + \omega = \omega.$$

The multiplication $\alpha \cdot \beta$ is realised by replacing each element of β with a copy of the entire well-ordered set α. The exponentiation is even more complicated.

In this way, an exotic ordinal number arithmetic is created, which neither fulfils commutativity nor usual cancellation rules. Also, fascinating ordinal numbers can be defined, such as the number ε_0, which is given as an unfathomably large power

$$\varepsilon_0 = \omega \uparrow \omega = \underbrace{\omega^{\omega^{\cdot^{\cdot^{\cdot^\omega}}}}}_{\omega \text{ times}}.$$

This number fulfils the unusual fixed point equation

$$\omega^{\varepsilon_0} = \varepsilon_0.$$

Due to the well-ordering of ε_0, however, every strictly descending sequence of ordinal numbers

$$\varepsilon_0 > \alpha_1 > \alpha_2 > \cdots$$

is necessarily finite. Ordinal numbers are the basis for the incredibly effective proof technique of transfinite induction, in which—as with complete induction—the series of ordinal numbers is climbed.

All ordinal numbers $\alpha < \varepsilon_0$ can be associated with finite trees that have a distinguished vertex—the root. Trees are graphs that do not contain multiple edges and no loops, i.e., closed paths that return to the starting point. A tree with a root and attached subtrees $B_1, \ldots, B_m$ corresponds to the ordinal number in Cantor's normal form

$$\alpha = \omega^{\beta_1} + \cdots + \omega^{\beta_m},$$

where $\beta_1 \geq \cdots \geq \beta_m$ are ordinal numbers that correspond to the subtrees. In the figure, we have indicated some examples.[114]

John von Neumann defined around 1928 the cumulative hierarchy of sets

$$V_0 = \emptyset$$
$$V_{\alpha+1} = \text{Pow}(V_\alpha)$$
$$V_\lambda = \bigcup_{\alpha < \lambda} V_\alpha,$$

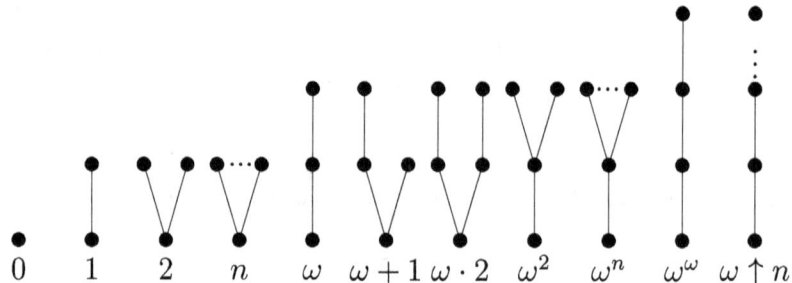

$$0 \quad 1 \quad 2 \quad n \quad \omega \quad \omega+1 \quad \omega \cdot 2 \quad \omega^2 \quad \omega^n \quad \omega^\omega \quad \omega \uparrow n$$

Ordinal numbers and root trees (each at the bottom).

which is based on the empty set. Here, α is any ordinal number and λ is a limit ordinal number, i.e., it is not of the form $\lambda = \lambda' + 1$ for an ordinal number λ'. The smallest limit ordinal number is obviously ω. For very large ordinal numbers κ, which are called strongly inaccessible cardinal numbers, the sets V_κ of the cumulative hierarchy are also called Grothendieck universes and are models of the Zermelo–Fraenkel axioms, as can be shown by transfinite induction. The postulation of strongly inaccessible cardinal numbers represents one of many possible additional axioms, which imply the consistency of the Zermelo–Fraenkel axioms.[115]

In his paradise, Cantor formulated the continuum hypothesis. It states that there is no set whose cardinality is strictly between the cardinality $\aleph_0$ of the natural numbers and the cardinality $2^{\aleph_0}$ of the real numbers. This conjecture cannot be proven because there are models of set theory, which satisfy the continuum hypothesis, as Kurt Gödel proved with the help of the cumulative hierarchy, and other models that do not satisfy it. The latter was shown by Paul Cohen with the help of the forcing method. In this method, new sets are constructed from a given and sufficiently small model of set theory, by the refined trick of adjunction, which also occurs in algebra.[116]

Interestingly, Skolem had anticipated the idea of the forcing method in set theory long before. He wrote in a paper from 1922:

> It would certainly be of much greater interest if one could prove that a new subset Z_0 could be adjoined without contradictions arising; but this will probably be very difficult.[117]

Free Objects and Adjunction

The method of adjunction is very common in mathematics and consists of adding new independent generating elements to a given structure, so that certain properties are preserved, but the new elements acquire the desired properties in a controlled manner.

The adjunction method consists of two steps. In the first step, a generic new element is added, which is independent of the elements of the initial situation. In this way, a free structure is created. In the second step, conditions in the form of relations are imposed on the new elements, so that the calculation rules of the old elements are preserved and the newly constructed object fulfils the desired rules.

A beautiful example of a free object are the natural numbers $\mathbb{N}$. They form a free monoid[118] with addition as the operation and a single generating element 1, because every natural number is a sum of values of the number 1:

$$5 = 1 + 1 + 1 + 1 + 1.$$

The number 0 also falls under this definition, as it arises when the number 1 is summed up zero times.

Free monoids can have more than one generating element. The free monoid over two elements consists—written multiplicatively—of finite words in the letters a and b. Such words can be graphically visualised as nodes in a binary tree (see figure).

Similarly, there are free groups[119] over one or more generating elements. The free group with a single generating element is the set of integers $\mathbb{Z}$. Free groups with two or more generating elements are no longer commutative and are therefore preferably noted as multiplicative groups, as the symbol $+$ is usually reserved for commutative groups.

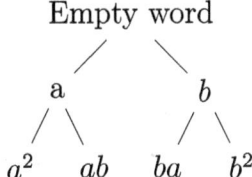

Empty word

$$a \qquad b$$

$$a^2 \quad ab \quad ba \quad b^2$$

Binary tree for words in two letters.

The free group F_2 over two generating elements a and b consists of arbitrary finite words in the two letters and their inverses a^{-1} and b^{-1}, such as the word

$$w = aba^{-1}b^{-1}.$$

Free objects possess a characteristic universal property. This states for F_2 that for every group G with two generating elements, whether commutative or not, there is a mapping

$$F_2 \longrightarrow G$$

which maps the two generating elements of F_2 onto the generating elements of G. This mapping is surjective, i.e., it reaches every element of G, but it is generally not injective, unless G is itself a free group. A similar universal property applies for free monoids and for free commutative groups $\mathbb{Z}^n$ of rank n.

With this knowledge, we now want to look at a concrete example of the adjunction method. Given a ring R, there is the possibility to add a new element T to add to the object R, and to form the free object

$$R[T],$$

which is called the polynomial ring over R. The elements in $R[T]$ are polynomials, i.e., finite sums

$$f = a_d T^d + \cdots + a_2 T^2 + a_1 T + a_0,$$

where the coefficients a_i are elements of R and the powers T^i are new expressions. All powers form infinitely many new elements and the finite sums of such expressions form an infinitely large set. The polynomial ring $R[T]$ is therefore a much larger object than R itself. The symbol T is usually called an unknown or variable. The addition and multiplication in this new object $R[T]$ is explained in an obvious way.

The polynomial ring $R[T]$ is a free structure, because no conditions were imposed on the variable T and the powers T^i form a copy of the natural numbers, because the exponent i can exactly run through the natural numbers.

In the second step of the adjunction, a relation $f = 0$ is imposed. This constructs a new object, denoted by

$$S = R[T]/(f).$$

In S, two polynomials from $R[T]$ are identified when their difference is a multiple of f. This procedure thus enforces the equation $f = 0$.

As an example, we take the polynomial $f = T^2 - 2$ and form the object

$$S = R[T]/(T^2 - 2).$$

What corresponds to the element T in this construct? Since we have enforced the equation

$$T^2 - 2 = 0$$

with the relation, it holds for the image of the variable T in S that $T^2 = 2$ is, i.e., T is a square root of 2:

$$T = \sqrt{2}.$$

Through two steps, we have constructed a new ring S from the ring R in which the element 2 has a square root, even though it did not necessarily have a square root in R before. We write this as

$$S = R[\sqrt{2}].$$

The ring S is different from R exactly when $\sqrt{2}$ did not exist in R. In the case $R = \mathbb{Z}$, this is obviously the case.

Another example is given by the polynomial $f = T^2$. In this case,

$$S = R[T]/(T^2)$$

is the ring of infinitesimal numbers over the ring R, which was used by Leibniz in his infinitesimal calculus. The element T in this ring can be interpreted as an infinitesimal object.

Geometry and the Concept of Space

Geometric shapes model many situations in applications of mathematics. We first want to consider as simple geometric structures as possible. The simplest 0-dimensional geometric object is obviously the point, which we denote by Δ_0 for our following purposes. In dimension 1 there is the unit interval $[0, 1]$, which we denote by Δ_1. The circle S^1 arises from the interval Δ_1, when the starting and ending point are identified. In this respect, Δ_1 is the prototypical geometric object of dimension one. In two dimensions there is the triangle Δ_2, the unit square $\square_2$ and many other polygons. By triangulation, all these objects can be decomposed into sums of the simplest triangles Δ_2. For example, the unit square $\square_2$ is made into the sum of two triangles by inserting a diagonal.

In higher dimensions it becomes more complicated. Already in antiquity, the 5 Platonic and the 18 Archimedean bodies were found. The Platonic bodies are, besides the cube $\square_3$ and the tetrahedron Δ_3, the octahedron, the dodecahedron and the icosahedron. In addition to Archimedean solids, there are also truncated bodies, such as the truncated icosahedron or football body, which consists of 5- and 6-sided polygons and has been used as a model for the football since 1970. From dimension 4, not all of these geometric shapes generalise and only the three series Δ_n (the n-simplex or hypersimplex), $\square_n$ (the hypercube) and $\Diamond_n$ (the cross polytope, the generalisation of the octahedron) remain. Of these three series, Δ_n is the most common and both $\square_n$ and $\Diamond_n$ can be decomposed into simplices Δ_n. It is practical to coordinate Δ_n by $n + 1$ coordinates $t_0, \ldots, t_n$ that satisfy the equations

$$0 \le t_i \le 1 \text{ and } t_0 + \cdots + t_n = 1.$$

The boundary $\partial \Delta_n$ of a single simplex Δ_n is given by the union of all parts of the form Δ_{n-1} that arise by setting a coordinate t_i to zero.

All the spaces we have considered so far are vivid elementary examples of topological spaces. A general topological space in mathematics is usually described as a set of points in which certain subsets are distinguished as open sets.[120] A useful class of topological spaces are the metric spaces, on which a concept of distance exists between any two points. The three-dimensional Euclidean space surrounding us $\mathbb{R}^3$ defines for two points $a = (a_1, a_2, a_3)$ and $b = (b_1, b_2, b_3)$ the Euclidean distance[121]

$$d(a, b) = \sqrt{(a_1 - b_1)^2 + (a_2 - b_2)^2 + (a_3 - b_3)^2}.$$

However, in mathematics there are numerous examples of topological spaces that are quite different from Euclidean space. For example, there are the non-euclidean geometries, discovered in the 18th century by János Bolyai, Nikolai Lobachevsky and Carl Friedrich Gauss.[122] In these geometries, Euclid's parallel postulate does not hold. With the help of differential geometry, which Bernhard Riemann subsequently developed, such spaces could be realised in particular as hyperbolic manifolds with constant negative curvature, in which the straight lines are given as geodesics, i.e., shortest connections between points. Manifolds are—figuratively speaking—geometric spaces with good smoothness properties that locally look like subsets of $\mathbb{R}^n$. In particular, a tangent space can be defined at each point.

The structure-preserving mappings between topological spaces are the continuous mappings. They are defined by the fact that they preserve the concept of distance in a certain way. This is equivalent to the fact that for any two points $a, b \in X$ that are close enough together, the image points $f(a)$, $f(b)$ are not far apart.[123] Two spaces have an indistinguishable topological structure—i.e., they are homeomorphic—when there is a bijective continuous mapping f between them, such that both f itself and the inverse mapping of f are continuous.

Symmetries

Topological spaces often possess symmetries. For example, the three-dimensional Euclidean space $\mathbb{R}^3$ that surrounds us has a symmetry group consisting of translations and rotations that preserve lengths and angles. Anyone who jumps from a diving board in a swimming pool utilises these possibilities of movement. The Platonic and Archimedean solids and many ornaments in artworks and buildings since antiquity possess finite symmetries. Symmetry groups in physics are closely related to conservation quantities according to a theorem by Emmy Noether.[124]

A symmetry group G operates on a topological space X by each group element g forming a homeomorphism $g \colon X \longrightarrow X$. The sets $\{gx \mid g \in G\}$ for a fixed point $x \in X$ are referred to as orbits and there is a continuous mapping $\pi \colon X \longrightarrow X/G$ from X to the orbit space X/G. An example is the real number line $X = \mathbb{R}$ and the additive group $G = \mathbb{Z}$, which operates as a symmetry group through translation. The orbit of each point x is given by all translates $x + n$ with $n \in \mathbb{Z}$. Through the quotient mapping $\pi(x) = \exp(2\pi i x)$, the orbit space $\mathbb{R}/\mathbb{Z}$ is identified with the unit circle S^1.

There are conditions for so-called good group operations on manifolds X, so that X/G still remains a manifold and π is an unbranched covering. The latter means that all orbits are isomorphic to G and for the mapping π for each $x \in X$ there exists a sufficiently small open set U in X, so that π forms a homeomorphism from U to $\pi(U)$. For less good operations, the quotients X/G form a class of spaces under certain conditions on the operation of G, which are called orbifolds or stacks. Individual points in X/G possess automorphisms as additional information and can become singular. A simple example of this is the operation of the group $G = \mu_n$ of the complex n-th roots of unity on $X = \mathbb{C}$. Then X/G is again homeomorphic to $\mathbb{C}$, it holds $\pi(z) = z^n$ and the zero point possesses the entire group μ_n as an automorphism group. Outside of the zero point, the operation is good.

Simplicial Spaces

One way to go from a geometric object to a simpler structure is the method of surveying, i.e., triangulation. Here, support points are chosen, connection curves between them, surface pieces between the connection lines, and so on up to higher dimensions. In the illustrated figure, the triangulation of a sphere S^2 is indicated.

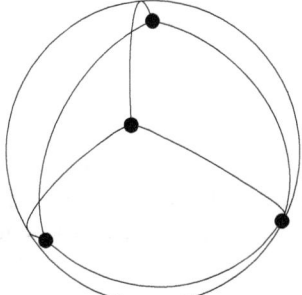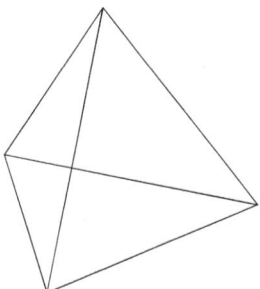

Triangulated sphere and its simplicial abstraction.

A triangulation of a topological space is an attempt to break it down into the simplest components, which are given by continuous mappings $f: \Delta_n \longrightarrow X$. The resulting combinatorial object, which is glued together from simplices, is called a simplicial space. Popular geometric bodies, such as spheres S^n, can be easily assembled in a combinatorial way from two or more copies of simplices Δ_n that are glued along the edges $\partial\Delta_n$.

Graphs are particularly simple simplicial spaces, consisting only of 0- and 1-dimensional simplices, i.e., from nodes (vertices) and edges. They form important structures in mathematics and are as fundamental as numbers. Graphs often describe combinatorial relationships. Relationships between people, organisations and objects in the world can be described using graphs. The simplest graphs are the trees. They contain no multiple edges and no loops, i.e., closed paths, that return to the starting point.

Simplicial spaces are special examples of geometric realisations of simplicial sets, the totality of which is denoted by **sSet**. These consist of an abstract set of simplices in each degree n and from gluing data between the simplices. Simplicial sets can be complicated and contain many simplices. Thus all continuous mappings $f: \Delta_n \longrightarrow X$ for a topological space X can be considered and we obtain a simplicial set, which is denoted by $\mathrm{Sing}_\bullet(X)$ and plays an important role. Simplicial sets can be seen as combinatorial-algebraic abstraction of topological spaces. Their geometric realisations possess good properties.[125]

Paths, Fundamental Group and Homotopies

The actual shape of topological spaces and their generalisations is not decisive in many contexts, but rather the equivalence class, which arises when continuous deformations, called homotopies, are allowed. Their equivalence classes are called homotopy types.

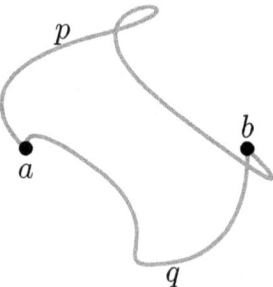

Two paths p, q from a to b.

What is a homotopy? We will first explain paths in topological spaces and then expand the concept. A path p in a topological space X is given by a continuous mapping

$$p \colon [0, 1] \longrightarrow X,$$

where [0, 1] is the real unit interval. Here, $p(0) = a$ is the starting point and $p(1) = b$ is the end point of the path (see figure).

Paths p and q can naturally be linked to a path $q \circ p$ when the end point of p coincides with the starting point of q (see figure). Expressed in mathematical language,

$$(q \circ p)(t) = \begin{cases} p(2t) & t < \frac{1}{2} \\ b = p(1) = q(0) & t = \frac{1}{2} \\ q(2t - 1) & t > \frac{1}{2}. \end{cases}$$

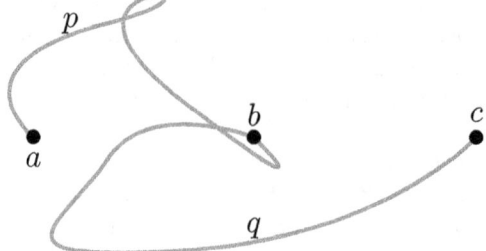

Linking $q \circ p$ of two paths.

The inverse path p^{-1} to a path p from a to b is a path from b to a and is given by reversing the direction of path p. It is given by the function $p^{-1}(t) = p(1-t)$, where t is the coordinate in the interval $[0, 1]$.

Two paths p, q with the same starting and ending points are homotopic to each other, denoted $p \simeq q$, when there is a homotopy between them. Such is given by a continuous mapping

$$h: [0, 1] \times [0, 1] \longrightarrow X,$$

so that $p(t) = h(0, t)$ and $q(t) = h(1, t)$ define the two paths (see figure). For the defined operations, the following calculation rules apply up to homotopy:

$$p^{-1} \circ p \simeq 1_a$$
$$p \circ p^{-1} \simeq 1_b$$
$$(p \circ q) \circ r \simeq p \circ (q \circ r).$$

Here, 1_a and 1_b are the constant paths at a and b. The homotopy of paths in these formulas is based on a reparametrisation of the paths as mappings from the unit interval to X. The formulas become incorrect if the symbol $\simeq$ is replaced by equality, because already the first formula is only valid up to homotopy.

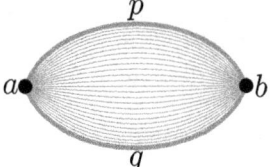

A homotopy between two paths p and q from a to b.

The fundamental group $\pi_1(X, *)$ is the group of homotopy classes of all paths in X with fixed start and end point $*$, the base point. It is usually not commutative, as the example of the "lying figure eight" (infinity symbol) shows, whose fundamental group is the free group F_2 over two generating elements, which correspond to the two obvious paths on the depicted figure.

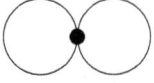

The "lying figure eight".

The fundamental group has some disadvantages. It obscures individual properties of paths in X and the dependence on the base point is cumbersome through the homotopies. Moreover, there are interesting topological spaces that hardly allow real paths and for which our definition therefore provides nothing. For example, the usual definition is usually worthless in algebraic geometry. In this case, it is better to define the fundamental group using so-called étale coverings of X, which are particularly unbranched. The standard definition is only suitable for topological spaces that have enough real paths.

Often it is better to consider paths between all points, without necessarily limiting oneself to homotopy classes and a base point. In the general case, we therefore work best with a higher categorical structure $\Pi_\infty(X)$, which is called the fundamental infinity groupoid or more generally infinity groupoid. In it, all calculation rules only apply up to certain equivalences, i.e., homotopies and higher generalisations thereof.

What is a general homotopy and a homotopy equivalence? For this, we consider two continuous mappings

$$f, g \colon X \longrightarrow Y$$

and call both homotopic, if there is a continuous mapping

$$h \colon [0, 1] \times X \longrightarrow Y$$

with $h(0, -) = f$ and $h(1, -) = g$. Two spaces X, Y are then called homotopy equivalent, if there are continuous mappings

$$F \colon X \longrightarrow Y \text{ and } G \colon Y \longrightarrow X,$$

so that $G \circ F$ and $F \circ G$ are each homotopic to the identity on X and Y. There is a somewhat more general concept of weak homotopy equivalence between two spaces. They preserve essential characteristics and invariants of topological spaces.

Topological Invariants

Properties of equivalent objects in mathematics, which are common to all representatives, are called invariants. In mathematics and physics, there are many invariants that carry crucial information about the objects under consideration. Invariants of topological spaces play a special role, as they are preserved under homeomorphism or under more general operations like homotopy equivalence.

The mathematician Emmy Noether has made a decisive contribution to ensuring that invariants of algebraic, topological and physical nature were assigned their appropriate significance. She defined homological invariants of abstract algebraic chain complexes and thus initiated the algebraisation of topology.[126]

How does this work? Each topological space can be assigned a natural simplicial set and thus a natural singular chain complex.[127] For this, we consider the set of all continuous mappings of the simplices Δ_n to X:

$$\mathrm{Sing}_n(X) = \{f \colon \Delta_n \longrightarrow X \text{ continuous}\}.$$

This defines a canonical simplicial set, which is denoted by

$$\mathrm{Sing}_\bullet(X).$$

It also includes boundary mappings

$$\partial_n : \mathrm{Sing}_n(X) \longrightarrow \mathrm{Sing}_{n-1}(X),$$

which are given by restriction to the boundaries $\partial \Delta_n$ of the simplices. The elements of $\mathrm{Sing}_n(X)$ are called n-simplices, just like the objects Δ_n themselves. The n-simplices degenerate when n is larger than the dimension of X. They are omitted for many purposes, but not always, because the simplicial set $\mathrm{Sing}_\bullet(X)$ contains valuable homotopy-theoretical information about X, which is then lost.

To every simplicial set $S_\bullet$—and therefore to every topological space X through the simplicial set $\mathrm{Sing}_\bullet(X)$—homology groups can be assigned, as Emmy Noether had introduced in general. In this view, we consider the chain complex of abelian groups underlying the simplicial set $S_\bullet$

$$\mathbb{Z}S_\bullet : \cdots \longrightarrow \mathbb{Z}S_n \longrightarrow \mathbb{Z}S_{n-1} \longrightarrow \cdots \longrightarrow \mathbb{Z}S_1 \longrightarrow \mathbb{Z}S_0,$$

where $\mathbb{Z}S_i$ is the free abelian group generated by the simplices in S_i. The linear mappings between these free groups are generated by the given boundary mappings $\partial_i : S_i \to S_{i-1}$ in $S_\bullet$. Then we form at each point the homology groups $H_n(S_\bullet)$ as the quotients

$$H_n(S_\bullet) = \frac{\mathrm{Ker}(\mathbb{Z}S_n \longrightarrow \mathbb{Z}S_{n-1})}{\mathrm{Im}(\mathbb{Z}S_{n+1} \longrightarrow \mathbb{Z}S_n)}.$$

The n-th Betti number b_n is then defined as the rank of $H_n(S_\bullet)$. The zeroth Betti number b_0 gives the number of connected components of $S_\bullet$ and the first Betti number b_1 the number of loops. The n-th Betti number indicates whether a higher-dimensional "hole" exists.

We want to give some simple applications of this formal algebraic definition. First, let's consider a triangle, i.e., the boundary of the 2-simplex (see figure).

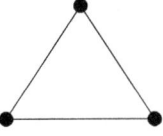

Triangle $\partial \Delta_2$.

The triangle consists of three copies of Δ_1 as edges and three vertices Δ_0. We give all edges a counter-clockwise orientation. This defines the end and the starting point of each edge. The chain complex $S_\bullet$ which calculates the homology groups is of the form

$$\mathbb{Z}^3 \xrightarrow{\partial} \mathbb{Z}^3,$$

where ∂ calculates "end point minus starting point" from each edge and is represented in a suitable basis by the matrix

$$\begin{pmatrix} -1 & 0 & 1 \\ 1 & -1 & 0 \\ 0 & 1 & -1 \end{pmatrix}.$$

The entries $+1$ stand for the endpoints and -1 for the starting points. This matrix has determinant 0 and rank 2, because the first two columns are linearly independent and the sum of the columns or rows is 0. From this it follows that both the kernel of ∂ and the cokernel of ∂ are isomorphic to $\mathbb{Z}$. Therefore, the Betti numbers are $b_0 = b_1 = 1$.

A slightly different example is the already mentioned "lying figure eight", where two loops Δ_1 and only one point $\Delta = 0$ occur. In this figure, the chain complex is given by

$$\mathbb{Z}^2 \xrightarrow{\ \partial\ } \mathbb{Z},$$

where ∂ is given by the zero mapping, because the end and starting point of the 1-simplices coincide. Thus $b_1 = 2$ and $b_0 = 1$ hold. More generally, in graphs, the Betti number b_1 is equal to the number of loops.

The higher-dimensional "holes" are given by the higher Betti numbers b_n. For spheres, for example, it holds

$$b_i(S^n) = \begin{cases} 1 & \text{for } i = 0 \text{ and } i = n \\ 0 & \text{otherwise.} \end{cases}$$

The homotopy groups $\pi_n(X)$ of a topological space X are defined as homotopy classes of continuous mappings

$$f : S^n \longrightarrow X.$$

A suitable composition can be defined on the homotopy groups, so that they carry a group structure and for $n \geq 2$ are even commutative. The most important of these groups is the fundamental group $\pi_1(X, *)$, where $*$ is a base point in X. A space X is called a homotopy n-type, if the homotopy groups $\pi_i(X)$ for $i > n$ are zero.

A famous example of a homotopy group is provided by the 2-sphere S^2, for which there is the Hopf fibration

$$S^3 \longrightarrow S^2$$

which contains exactly the crucial information about the generator of the third homotopy group $\pi_3(S^2) \cong \mathbb{Z}$. The sphere S^2 is therefore not a homotopy 2-type.

Mathematics is a multicultural science with a rich past. We do not know exactly when and where it first originated. Beginnings that go beyond simple counting methods can be found in the ancient Orient long before Greek antiquity, among the Maya, in China and India, and in many other places in the world. It was certainly a cultural technique from the outset, which had concrete applications in practical areas of the societies of the time. Only in Greek antiquity did a mathematical science slowly develop with the fundamental areas of arithmetic, geometry and logic.[128]

Most people are only aware of a few applications of mathematics. In recent times, blockchain algorithms, cryptographic protocols and machine learning have gained a high level of recognition due to digitalisation. Numerous other examples can be found in technology and natural sciences as well as in the social sciences.

Contrary to widespread beliefs, however, mathematics is not limited to quantitative and algorithmic methods. Its actual task is to provide concepts and buildings for other sciences. The interlocking with physics is remarkable, as we have already indicated in the introduction. In the future, further challenges await mathematics. One of them comes from the exploration of complex systems, where multidisciplinary phenomena need to be understood and managed.

Cryptography or the Art of Encryption

Encryption methods are used to exchange secrets securely or to generate tamper-proof certificates and authentications.[129] These protocols often calculate in residue classes modulo a very large number N, i.e., with the numbers

$$\mathbb{Z}/N\mathbb{Z} = \{0, 1, 2, 3, \ldots, N-1\}$$

S. Müller-Stach, *The Code of Mathematics*, Mathematics Study Resources 11,
https://doi.org/10.1007/978-3-662-69483-1_4

and these residue classes are added and multiplied in the usual way, with multiples of N possibly being subtracted if the result is strictly greater than $N - 1$. This is familiar to most people when $N = 7$ is, because the seven days of the week

$$\text{Mon, Tue, Wed, Thu, Fri, Sat, Sun}$$

form residue classes modulo $N = 7$.

In cryptography or when entering passwords into computer systems, one-way functions are used, which are often also called hash functions. With these, it is possible to verify the confidentiality or the verification of the truth of statements—such as the authentication of a person—without jeopardising security and anonymity. With mathematical methods, many hash functions can be generated and confidentiality protocols can be technologically realised and made secure. This is where the actual value of number theory in cryptography lies.

Hash functions are functions with the property that from a function value $f(x)$ the argument x can only be calculated with great difficulty. They are called one-way functions when the corresponding function f is injective, i.e., when different arguments have different function values. Hash functions are comparable to classical or genetic fingerprints, as they possess similar properties. The powers

$$g^m \bmod N$$

form the most important basis of elementary mathematical hash functions and are used in RSA encryption, ElGamal encryption, in Diffie–Hellman key exchange and in Shamir's three-pass protocol. Here we can either consider the base g or the exponent m as a secret.

The calculation of m given the power g^m of g is called the discrete logarithm problem. Neither trying out all possibilities nor index calculations or the sophisticated algorithms of Pohlig–Hellman, Pollard and Shanks significantly alleviate this difficulty. The security of ElGamal encryption and the Diffie–Hellman key exchange rely on this.

In many protocols, $N = p$ is a large prime number. In practice, often only the elements in $\mathbb{Z}/N\mathbb{Z}$ are used that are coprime to N. These form a multiplicative subgroup U_N of units in $\mathbb{Z}/N\mathbb{Z}$. In the case of RSA encryption, which we will now go into in a little more detail, $N = pq$ is the product of two different large prime numbers with more than a hundred digits.[130]

The Security of RSA Encryption

In the case of RSA encryption, the secret is given by a number $g \in \mathbb{Z}/N\mathbb{Z}$. The security of the procedure is guaranteed if it is difficult for large N to determine the base g from the knowledge of g^m, even if m is known.

In the case $N = p$ prime, there is a simple decryption method for this problem, called the backdoor. For this, we search with the help of the Euclidean algorithm for a m' with

$$mm' \equiv 1 \bmod p - 1$$

and Fermat's little theorem[131] guarantees that

$$(g^m)^{m'} \equiv g^{mm'} \equiv g \bmod N.$$

The decryption is therefore also carried out with a simple exponential function. This is obviously not a particularly secure method if m is known, because the number m' is easy to calculate from m if $N = p$ is a prime number.

This behaves quite differently when $N = pq$ is the product of two large different prime numbers. The decryption g of g^m given exponent m is also obtained in this case by exponentiation with a suitable m', if the factorisation of $N = pq$ is known. This time, m' is not so easy to determine, but requires Euler's theorem, which generalises Fermat's little theorem. This theorem states that

$$g^{\varphi(N)} \equiv 1 \bmod N,$$

if g is coprime to N, where $\varphi(N) = (p - 1)(q - 1)$ is the Euler φ-function[132] of N. The decryption (or backdoor) is then given by exponentiation with m', where

$$mm' \equiv 1 \bmod \varphi(N).$$

RSA encryption is therefore secure because it is just as difficult to factorise a large number N of the form $N = pq$ as it is to calculate $\varphi(N) = (p - 1)(q - 1)$. A malicious person can, even if they know N, only calculate the number m' and thus g from the knowledge of g^m and m with unreasonable time and computational effort.

Factoring large natural numbers poses a problem, which is presumably more difficult than verifying whether a number is prime. In 2002, it was shown that prime number tests in polynomial time are possible and for this the expression "Primes is in P" was coined. For the factorisation of numbers, there are algorithms, such as the quadratic sieve or the number field sieve, which have subexponential complexity. However, no polynomial time algorithm is known. Peter Shor proved in 1994 that a quantum computer would achieve this in polynomial time.[133] Because of this fact, cryptographic postquantum protocols for the future are being researched.

A Protocol for Exchanging Secrets

In Shamir's three-pass protocol, two people, usually referred to as Alice and Bob in cryptography, want to exchange a secret without a malicious person, often called Eve (from English evil), being able to uncover it. The idea can be easily explained without mathematics. The concept is that the secrets are sent via a regular, lockable suitcase, to which Alice and Bob can attach locks with their own keys. Alice puts the secret in the suitcase and locks it with her own key, which no one else possesses, not even Bob and certainly not a malicious Eve. Alice hands the suitcase to Bob, who in turn locks the suitcase a second time and returns it to Alice. Now Alice has the opportunity to unlock her own lock without the suitcase being able to be opened, because Bob's lock is still attached. Finally, she gives the suitcase back to Bob, who can now unlock it and share the secret with Alice. During this process, Eve never had the opportunity to find an unlocked suitcase.

If Alice and Bob's locks are realised with natural numbers, Alice chooses private numbers m and m' with

$$mm' \equiv 1 \bmod \ p - 1$$

and Bob his private numbers n and n' with

$$nn' \equiv 1 \bmod \ p - 1.$$

Then the protocol is realised by Alice first sending the number g^m to Bob, then Bob forming the power $(g^m)^n = g^{mn}$ and sending it back to Alice, who in turn forms the m'-th power:

$$(g^{mn})^{m'} = g^{mm'n} \equiv g^n \bmod \ p.$$

Finally, Bob exponentiates this number again with the n'-th power and obtains

$$(g^n)^{n'} = g^{nn'} \equiv g \bmod \ p.$$

Alice has thus successfully sent the secret g to Bob. Along the way, the numbers g^m, g^{mn} and g^n appear, from which g cannot be obtained with realistic effort without essentially trying out all possible powers.

Similar methods can be used to realise digital signatures, i.e., signatures and other authenticity certificates, and enable zero-knowledge proofs, with which the truth of statements can be checked without seeing the contents.

Blockchains

The fascinating new technology of blockchains is also based on hash functions. Commercial hash functions such as the secure-hash algorithm SHA 256 are used, which differ from the hash functions we have considered.

Blockchains were invented by Satoshi Nakamoto. This name is a pseudonym and it is not known who is behind it. In the only work under this name, this person wrote:

> A purely peer-to-peer version of electronic cash would allow online payments to be sent directly from one party to another without going through a financial institution. Digital signatures provide part of the solution, but the main benefits are lost if a third party is still required to prevent double-spending. We propose a solution to the double-spending problem using a peer-to-peer network. The network timestamps transactions by hashing them into an ongoing chain of hash-based proof-of-work, forming a record that cannot be changed without redoing the proof-of-work.[134]

A blockchain is a list of blocks that represent confidential data and are gradually extended. The processing of blockchains takes place in communities of computer networks, whose nodes possess private identities. There is no central organisation, such as a bank, that monitors the proceedings. The appending of new blocks takes place in a mathematical consensus procedure in the form of a proof-of-work. The basic idea of blockchains is that each block is connected with the past blocks, i.e., the data of previous processes and the identities of the participants are preserved at each step in the form of values of hash functions and can be individually stored as evidence (see figure). The verification of new transactions in the proof-of-work is extremely computationally intensive. This requires a mathematical performance, which is called mining.

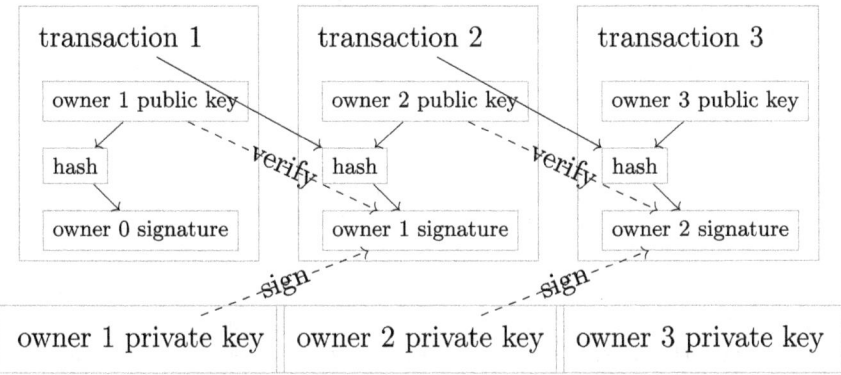

Flowchart of blockchains according to Satoshi Nakamoto.

An important application of blockchains are cryptocurrencies such as bitcoins. The total volume of all bitcoins is capped and the exchange rate is quite volatile. Often the question is asked, what is the equivalent value of bitcoins and other cryptocurrencies, as they ultimately only exist as data. The answer is like cash. Perhaps with a historical coin, the issue value still roughly corresponded to the metal value. At the latest, banknotes have broken with this tradition and the value of a carefully produced and difficult to counterfeit banknote is that it can be used to purchase other things without any problems. The value of money thus corresponds to the trust associated with it. The same is true for cryptocurrencies. Whether and in what form they will establish themselves in the future is still unclear, especially since the high energy consumption is extremely harmful to the environment.

Factorisation and Quantum Computing

Classic digital computers calculate with bits, which are elements of the set $\{0, 1\}$. This idea ultimately goes back to Leibniz. They calculate functions that map m-bit vectors to n-bit vectors:

$$f : \{0, 1\}^m \longrightarrow \{0, 1\}^n.$$

Such functions can be practically arbitrary. However, all conceivable calculations are realised using a few standard gates that mimic logical operations. The AND, OR and NOT gates form a generating set from which any other function f can be generated. These gates correspond to the logical operations $\wedge$, $\vee$ and $\neg$. The NAND gate, i.e., the negated AND gate, already generates these alone.

Quantum computers work with qubits instead of bits and utilise the laws of quantum mechanics. Physical states in quantum mechanics are vectors in a Hilbert space of finite or infinite dimension. The temporal dynamics of quantum mechanical processes are described by the Schrödinger equation. From the theory of quantum mechanics, it follows that the physically permissible operations f that can occur between such Hilbert spaces are either projection mappings or unitary operators. Projections correspond to measurements on a quantum mechanical system.

A qubit is an element in the two-dimensional Hilbert space $H = \mathbb{C}^2$ with the standard scalar product. The basis of H is often denoted in physics as

$$|0\rangle, |1\rangle.$$

A vector in H is given by a linear combination

$$\alpha|0\rangle + \beta|1\rangle,$$

where α and β are complex numbers. When n such qubits are considered together, the appropriate Hilbert space that describes the overall situation is the n-fold tensor product

$$H^{\otimes n} = \underbrace{H \otimes H \otimes \cdots \otimes H}_{n \text{ times}}.$$

This complex vector space has the dimension 2^n and every vector in H is of the form

$$u = \lambda_{00\ldots0}|00\ldots0\rangle + \cdots + \lambda_{11\ldots1}|11\ldots1\rangle.$$

This large dimension is the actual reason why quantum computers can be so powerful. A state u is called entangled if it cannot be written as a tensor product $v \otimes w$. The simplest case of this kind occurs at $n = 2$, because

$$u = |01\rangle + |10\rangle$$

is not an unentangled tensor product of the form

$$(\alpha|0\rangle + \beta|1\rangle) \otimes (\gamma|0\rangle + \delta|1\rangle).$$

This follows from the fact that the coefficients λ_{ij} of the unentangled vectors for $n = 2$ satisfy the cone equation

$$\lambda_{00} \cdot \lambda_{11} - \lambda_{01} \cdot \lambda_{10} = 0$$

but u does not. There are suitable physical experiments that produce entangled particles. For example, entangled pairs of photons can be generated via polarisation, which even at a great distance from each other retain this property at times. This leads to interesting transmission channels and cryptographic protocols in quantum information theory.

Quantum gates are the necessary gates used in the context of quantum computers. Unlike with digital computers, gates are necessary that induce unitary operations. Examples of such gates are the Hadamard gate H and the CNOT gate, i.e., the controlled NOT gate:

$$H = \frac{1}{\sqrt{2}} \begin{pmatrix} 1 & 1 \\ 1 & -1 \end{pmatrix}, \quad \text{CNOT} = \begin{pmatrix} 1 & 0 & 0 & 0 \\ 0 & 1 & 0 & 0 \\ 0 & 0 & 0 & 1 \\ 0 & 0 & 1 & 0 \end{pmatrix}.$$

In 1994, Peter Shor found the Shor algorithm for quantum computers, which is capable of factorising natural numbers in polynomial time, provided a quantum computer can calculate largely error-free with a sufficiently large number of qubits. Shor solved this task by implementing the so-called discrete Fourier transformation as a quantum algorithm and reading out the information, i.e., the coefficients in the base representation in $H^{\otimes n}$, in the final state of the system through a clever mix of quantum mechanical measurement and postprocessing using elementary number theory.[135]

Mathematical Physics

In his habilitation thesis, Bernhard Riemann developed the mathematical theory of manifolds together with their metric and differential geometric structure. This theory has particularly contributed to the discovery of Einstein's field equations in general relativity theory.

In mathematical physics, the concept of spacetime is usually modelled on Riemannian manifolds, where in cosmology singularities must also be allowed, which explain phenomena such as black holes. It is common to describe a manifold M and its metric-differential geometric structure locally using coordinates. However, global properties play a more decisive role. On M and its tensor fields often operates a group of gauge transformations as a symmetry group, which preserves crucial physical quantities. The physics to be described with such a differential geometric model does not depend on the specifically chosen coordinates and the gauge transformations. This property is called covariance and is a good example that the mathematical description of physical objects is only relevant up to isomorphism or equivalence.[136]

The idea of covariance goes in parts back to Leibniz, who in his correspondence with Samuel Clarke from the years 1715–1716 discussed the structure of physical space. Leibniz wrote in a letter dated February 25th 1716:

As far as I am concerned, I have emphasised more than once that I consider space as well as time to be purely relative, an order of coexistence, just as time is an order of their succession ... I have several proofs with which I can refute the imagination of those who consider space to be a substance or at least an absolute being.[137]

In the same letter, Leibniz considered the possibility of swapping the cardinal directions East and West in space and then wrote further in loc. cit.:

But if space is nothing more than this order or relationship and if it is nothing without the bodies but the possibility of placing them, then these two states, namely the one as it really is, and the other, the completely reversed assumed one, are in no way different from each other.

Leibniz thus linked his concept of space with the concept of equality and isomorphism. Albert Einstein formulated the principle of covariance as follows:

The general laws of nature are to be expressed by equations that are valid for all coordinate systems, i.e., they are covariant with respect to arbitrary substitutions (generally covariant).[138]

Einstein's field equations describe the general theory of relativity. The decisive matrix-valued equation among them describes the relationship of the curvature tensors with the mass ratios in space and reads:[139]

$$R_{\mu\nu} - \frac{1}{2}g_{\mu\nu}R + \Lambda g_{\mu\nu} = \frac{8\pi G}{c^4}T_{\mu\nu}.$$

It is noteworthy that the ideas of Riemann together with Einstein's theory of relativity found an application in GPS navigation only after more than 150 years. The location determination using the radio signals of several satellites would be significantly less precise if the special and general theory of relativity were not taken into account at the same time.

Some remarks from Riemann's habilitation thesis are also worth mentioning. In the last section he wrote:

The question about the validity of the assumptions of geometry in the infinitesimal is linked to the question of the inner reason for the mass ratios in space. With this question, which may still be counted as part of the doctrine of space, the above remark applies that in a discrete manifold the principle of mass ratios is already contained in the concept of this manifold, but in a continuous one it must come from elsewhere. It must therefore either be the case that the real underlying space forms a discrete manifold, or the reason for the mass ratios must be sought outside, in binding forces acting on it.[140]

These remarkable sentences show deep insights and question the continuous space-time models that prevail to this day. Roger Penrose has developed spin networks as a new idea for discrete versions from around 1971. They replace geometric triangulations with their dual graph (see figure). Another approach is non-commutative geometry, where C^*-algebras take the role of local functions and no underlying topological space X is available. Both concepts could prove useful for the unification of quantum field theory with the general theory of relativity into the theory of quantum gravity.[141]

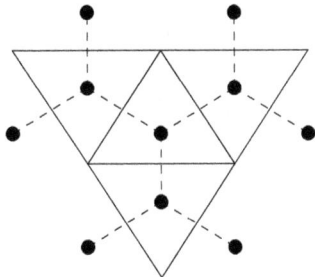

Spin network for a triangulation.

Similar to Einstein's field equation in general relativity, physical laws can often be described with one or more mathematical equations. In quantum mechanics, the Schrödinger equation

$$i\hbar\frac{\partial}{\partial t}\Psi(t) = H\,\Psi(t)$$

plays an important role for the wave function $\Psi(t)$. Another example is the heat conduction equation

$$\frac{\partial}{\partial t}u(x,t) - a\Delta u(x,t) = f(x,t).$$

It describes the propagation of temperature u in three-dimensional space (with the x-coordinate) depending on time t given a heat source $f(x,t)$ and conductivity index $a > 0$. Another example is the system of three Euler partial differential equations, which describe the flow of a fluid and their extension to the Navier–Stokes equations. The Clay Foundation in Boston named seven Millennium problems in 2000, the solutions of which pose challenges in mathematics and six of which have not been solved to this day. They are a reminiscence of the Hilbert problems from 1900. One of these problems asks about the existence and smoothness of solutions to the Navier–Stokes equations.[142]

Artificial Neural Networks and Deep Learning

Artificial intelligence is an umbrella term for many types of algorithmically controlled systems that simulate human thinking or can support and replace humans. The range is broad and includes suitable algorithms as well as machines, such as robots. After the beginnings of artificial intelligence in the times of Alan Turing and Norbert Wiener, the discovery of Hebb's rule by Donald Hebb in 1949 and the invention of the perceptron by Frank Rosenblatt in the late 1950s, numerous attempts over several decades have been made. During this time, statistical learning algorithms such as the PAC learning algorithm by Leslie Valiant, optimisation algorithms of various kinds, genetic and evolutionary algorithms and many others have been designed. Since around 2006, enormous breakthroughs have been achieved in a field called machine learning. A specific form of this includes deep learning, which involves training artificial neural networks.[143]

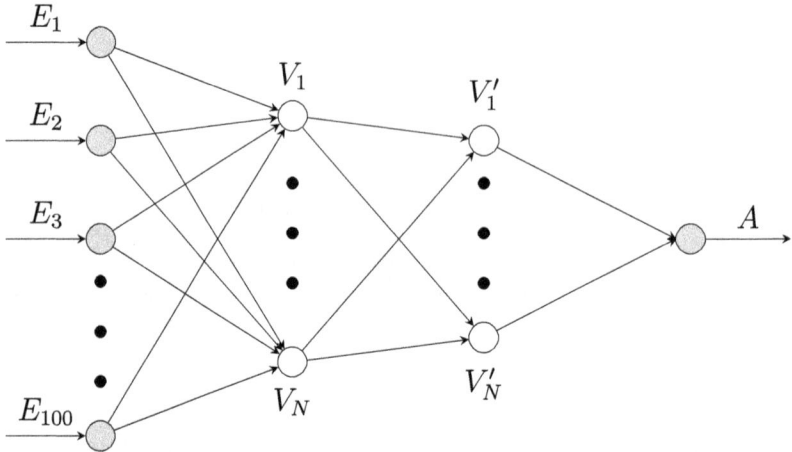

Artificial neural network with one output node.

Graph-like functional networks also occur in nature. Already in the 19th century, parts of the brain could be dissected so precisely that connected structures of neurons (nerve cells) and glial cells were discovered, with the neurons exchanging via synapses and being able to exhibit electrical activation. Around 1943, a simplified mathematical model of this system in the form of an artificial neural network was designed by Warren McCulloch and Walter Pitts.[144]

A simple form of an artificial neural network is the multilayer perceptron. It consists of a finite directed graph, which includes input nodes E. and output nodes A. as well as nodes V. from several hidden layers with their connecting edges. The number of hidden layers is referred to as the depth of the network and the number N its width. The figure suggests a blueprint for a network, with which from a digital square image with 10×10 pixels with grey values between 0 (white) and 1 (black) the letter Y is to be recognised with the help of a single output node.

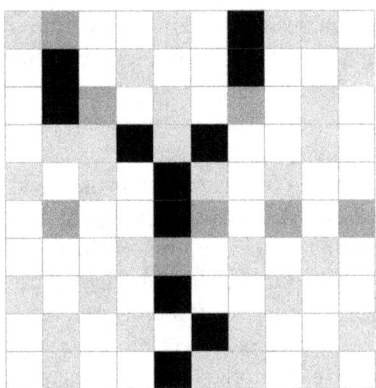

Training image "Y" with 10×10 grey pixels.

How is it calculated? The depicted neural network is intended to compute a real function $f \colon \mathbb{R}^{100} \longrightarrow \mathbb{R}$, whose arguments are the grey values e_i of the 100 pixels of a digital image. If the letter Y is recognised, then the network should approximate the function value $f(e_1, \ldots, e_{100}) \approx 1$ and otherwise yield $f(e_1, \ldots, e_{100}) \approx 0$.

Important for the proper functioning of a neural network are the transitions between the individual layers, also called feedforward. If we focus on the transition between the input nodes E. and a node V_i of the first hidden layer, this is realised for each i between 1 and N by formulas

$$v_i = \sigma(W_i + b_i),$$

where W_i are suitable functions of the input variables e_j, b_i are rational constants—called the bias or negative threshold—and σ is a non-linear activation function, for example of the form

$$\sigma(x) = \max(x, 0) = \begin{cases} 0 & \text{for } x \leq 0 \\ x & \text{for } x \geq 0. \end{cases}$$

In our example, W_i can be a simple linear function

$$W_i(e_1, \ldots, e_{100}) = w_{i,1} e_1 + \cdots + w_{i,100} e_{100},$$

where the coefficients $w_{i,j}$ are positive or negative rational numbers called weights. There are other choices for W_i and σ, which may be advantageous in some situations. Analogous formulas apply for the transitions in the hidden layers and in the final step to the output nodes.

How does the training of such a neural network work? At the beginning of the training, the weights $w_{i,j}$ and the (negative) threshold values b_i are initially set. The network is then trained by input data, by calculating for each input image a cost function K depending on the weights and threshold values. In the simplest case, K

measures the sum of the squares of the deviations from $f(e_1, \ldots, e_{100})$ from the desired value 0 or 1, which—in our example—is represented by the actual letter on the image. By successive adjustment of the weights $w_{i,j}$ and the threshold values b_i, the cost function K is slowly minimised. This method is called backpropagation or backtracking. Mathematically speaking, this process consists of a gradient descent, similar to the way of descending from a mountain as quickly as possible against the direction of the steepest slope. The goal is a local minimum of the cost function. To avoid nonsensical or unrealistic local minima of the cost function in this method, which do not come close to the solution of the problem, the gradient descent is usually combined with a stochastic component, which is randomly combined from several choices. Related to this is the method of genetic algorithms, with which unwanted local minima are avoided by evolutionary jumps in the parameters.

By inputting large amounts of data, the neural network can be adjusted over a long period of time until the weights and threshold values are altered so that it can recognise desired structural patterns from any new data with good accuracy. This makes the network a good predictor in appropriate new situations. Essentially, the method of deep learning involves reducing a high-dimensional problem to a few essential parameters, with the correct reduction found by training through approximation.

In addition to the simple case of the multilayer perceptron, there are more powerful recurrent neural networks, where certain directed edges transmit information backwards, thus enabling recursive calculations of sequential functions $f(n)$, which depend on a discrete parameter n. Other variants use sophisticated filters in the hidden layers. Such networks are called convolutional neural networks. They are particularly effective in image processing. Together with the enormous computing power of today's processors, such sophisticated refinements of the multilayer perceptron contribute to the success of artificial intelligence.

Pitfalls of Artificial Intelligence

A priori, there are many systematic errors in neural networks and other learning algorithms. Traditional results of analysis usually cannot accurately show the quality of the approximation, as the prerequisites of usual mathematical theorems, such as convexity, linearity or smoothness, are hardly ever met in the applications. Therefore, it is surprising that this method is so successful in many cases. The depth of the network, more than its width, plays a crucial role. At the moment, a new field of mathematical analysis of deep learning is emerging. On the one hand, it provides estimates of various types of systematic approximation errors and, on the other hand, it investigates fundamental problems and sources of error.[145]

It is noteworthy that deep learning, unlike other learning algorithms, is not significantly affected in its function by the phenomena of overfitting and overparameterisation. In the case of overfitting, the algorithm adapts optimally to the training data due to the numerous adjustable parameters, so that it can make perfect decisions in the given cases, but may not generalise well in many other situations and thus possibly produce incorrect results. The related overparameterisation is caused by the fact that many more adjustable parameters, such as the weights in neural networks, are present in the algorithm than the dimension of the essential parameters of the structures being examined.

Learning algorithms may produce solutions that are incorrect or have an undesirable bias, for example if the training data have a bias. A fundamental problem is that there are undecidable problems for deep learning, just as there are for ordinary Turing machines.[146] A danger comes from incorrect applications that arise from ignorance of the pitfalls of the methods. Questioning such algorithms is therefore justified. For these reasons, it is necessary to continue developing the theory quickly. The exploration of learning is a more general goal than the method of deep learning and includes other algorithms already discovered. The mathematical correctness of all these methods will one day presumably be verifiable and the use therefore safer and ethically less questionable than today.

Is artificial intelligence really a form of intelligence? This is a difficult question. Artificial intelligence in the form of deep learning is a method that can draw conclusions from complex problems with large amounts of data. This involves a reduction of complexity and an understanding in the form of predictors is achieved. Human intelligence is also capable of navigating complex situations and making good decisions. However, our intelligence seems to possess more properties than just the ability to reduce complexities in particularly complicated situations. The discovery of insights in the sciences, creative artistic work, or even understanding a subtle joke, are achievements of human intelligence that do not seem to be explainable. The last word has not yet been spoken. So there is room for further research.

Topological Data Analysis in Bioinformatics

The life sciences are increasingly applying methods from mathematics and computer science. The amount of data that can be generated from gene sequences of organisms, from molecular data, from image data in high-resolution imaging techniques, from configurations of neural networks in model organisms or from other methods is enormous and can only be managed with sophisticated methods. We want to address the problem of data analysis, which can be approached with mathematical methods from algebraic topology.

Mathematically, the aim is to recognise a geometric pattern from a large set of discrete data points in a high-dimensional space. Now, topological spaces and their invariants are precisely the objects that reflect typical geometric patterns. In applications, large collections of data points with a given distance concept, called metric, are relevant, which can have different manifestations. For genetic data, this could be the Hamming distance between vectors with combinatorial data of genetic sequences, which by definition is the number of all non-matching components. In neural networks, the reciprocal of the strength of the synaptic connections provides a metric.

There is a novel method to gain topological information from data point clouds. It is called persistent homology or topological data analysis. For this, the metric structure on the space of data is used to define a chain complex. There are two popular variants, firstly the Vietoris–Rips complex and secondly the Čech complex.

We first consider the Vietoris–Rips complex. The points P of the data cloud form the 0-simplices. Around each of these points P we place closed metric spheres $B_{P,\varepsilon}$ with radius $\varepsilon > 0$. Two different 0-simplices P and Q are connected by an edge if their distance is less than or equal to ε, i.e., if Q is contained in $B_{P,\varepsilon}$. Given pairwise different 0-simplices $P_0, \dots, P_k$, we define a k-simplex if all points have a distance less than or equal to ε from each other, i.e., if each point is in the closed ε-sphere of each other. In the case of the Čech complex, we define a k-simplex for $k + 1$ pairwise different points $P_0, \dots, P_k$, if the intersection

$$B_{P_0,\varepsilon} \cap \cdots \cap B_{P_k,\varepsilon} \neq \emptyset$$

is not the empty set. With both methods, we get a simplicial set $V(\varepsilon)$. for each ε. The associated chain complex $\mathbb{Z}V(\varepsilon)$. of abelian groups is called the Vietoris–Rips complex or Čech complex.

Persistent homology arises when we calculate the homology of the chain complexes $\mathbb{Z}V(\varepsilon)$. and investigate which homology groups and Betti numbers these chain complexes have when ε varies. These Betti numbers $b_n = b_n(\varepsilon)$ can be calculated using methods of linear algebra and can be visually represented in the form of barcodes depending on ε. The persistent homology is then the result of the barcodes that occur over longer distances (see figure). Mathematically speaking, the homology of filtered chain complexes, which are filtered over the parameter ε, is calculated in an efficient manner.

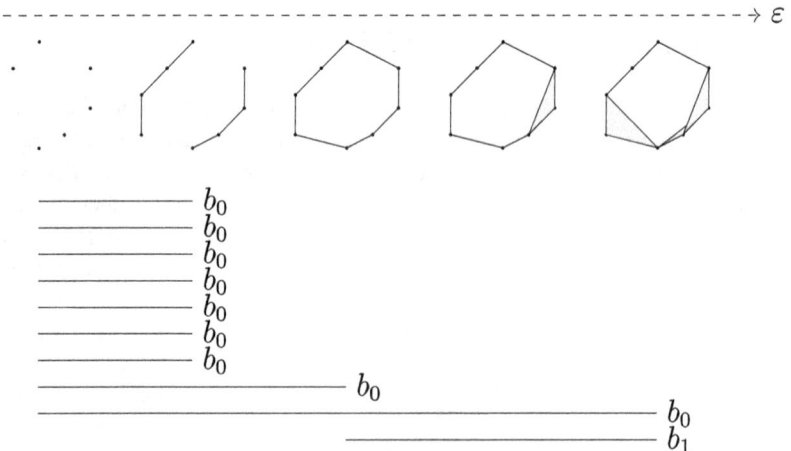

Barcodes of persistent simplices with increasing ε for a circular point cloud. The Betti numbers b_0 and b_1 stabilise from the third image to 1.

Topological data analysis has shown remarkable applications, particularly in the life sciences. However, it remains to be seen how successful it will actually be.

Opportunities for Mathematics

In the past, the sciences mostly worked in a monodisciplinary manner. Modern questions are often interdisciplinary and have complex systems as their mathematical basis. A good example of this is climate research and the Earth's ecosystem as a whole. This requires studying the physical and chemical foundations of meteorology and geosciences, as well as the influence of humans, which are interdependent. In the various layers of the Earth, physical interfaces spatially meet. Even if the dynamics occurring in each layer are understood, the spatial boundary transition must be modelled. This requires spatial multiscale methods and coarsening tricks as well as a high computational effort. In climate research, temporal multiscale methods are necessary to handle the dynamics on the time axis. While the entire climate operates on a large time scale, there are weather effects and human behaviour, both of which occur on a narrower time scale. For the mathematical explanation of the connection of these two scales, Klaus Hasselmann was awarded the Nobel Prize in 2021. His model demonstrated that the climate generally has a great stability, but in recent decades human-made effects outside this normal range of fluctuation have been added.

Another future task is the development of mathematical consensus protocols. Due to their immense computational effort and energy consumption, blockchains are no longer contemporary. Novel and more efficient data networks in the style of distributed ledgers are required, which are not monitored by any central authority. They are a special case of distributed systems, where many participants exchange data and execute algorithms without a central instance having a control function. Financial transactions in the areas of mobility, energy and health are good examples of this. The requirements for transparency, security and data protection are enormous. Complex and distributed systems are just two significant examples among others. Mathematics will play a crucial role in all future challenges, as it can design universally applicable models.[147]

The idea of recursion and complete induction is very old. But it was not until the 19th century that Richard Dedekind with his books[148] "Continuity and irrational numbers" and "What are and what should the numbers be?" established the fundamental properties of natural numbers on a solid foundation and defined the construction of the number system with integers, rational and real numbers from natural numbers. One of his most important achievements was the discovery and proof of the recursion theorem.

Building on this, a theory of computability only crystallised around 1936. At this time, almost simultaneously, all theoretical computability models were developed that we know today. They turned out to be equivalent and describe in an abstract fashion the strength of the later developed digital computers. In particular, the functions of $\mathbb{N}^n$ to $\mathbb{N}$ that can be computed with a Turing machine are exactly all partially defined recursive functions that can be constructed from elementary functions using primitive recursion and the μ-operator. From these contexts, the original Church–Turing thesis was derived, which informally states that every effectively computable function is recursive according to this definition. The question remains open whether there are stronger computability models that can be physically realised. To date, such expectations, referred to as hypercomputing, have not been realised, despite promising approaches for computer architectures in the field of neuromorphic computing.[149]

There are mathematical problems that cannot be solved algorithmically at all. These include in particular the halting problem for Turing machines, the word problem and the problem of solvability of integer diophantine equations. Such undecidable problems show us the limits of computability.

The Method of Recursion

Dedekind's recursion theorem states that a mapping f from the natural numbers to another set can be uniquely constructed by specifying $f(0)$ and providing a rule for how the value of $f(n)$ can be obtained from the values $f(m)$ for $m \leq n - 1$ for each $n \geq 1$.

We have already seen the method of recursion in the form of the Euclidean algorithm, where the crucial step is to reduce the problem to smaller numbers. Another example is the calculation of the factorial function, which is defined by the formula

$$f(n) = n! = 1 \cdot 2 \cdots (n-1) \cdot n.$$

For this function, $f(0) = 1$ and

$$f(n) = n \cdot f(n-1) \text{ for } n \geq 1.$$

The calculation of the function f at the argument n thus uses the product of n with the value of the function at the argument $n - 1$. Such a call of a value of a function at a smaller argument is called recursion. In most programming languages, recursive programs can be written so that the program code for the factorial function contains a similar line like

$$\text{return } n * f(n-1),$$

i.e., the function f calls itself. To better illustrate the principle of recursion, we want to explain the illustrative and at first glance unmathematical example of the Towers of Hanoi. There are three rods and n holed discs given, which are stacked on the first rod. The task is to move the discs of the tower one after the other so that they are stacked on the third rod at the end. The middle rod can be used as a help stack and a larger disc must never lie on a smaller one. There is a recursive solution strategy $L(n)$ for n discs. It consists of placing the top $n - 1$ discs on the help stack using the solution strategy $L(n-1)$, then placing the largest disc on the third rod and then applying the solution strategy $L(n-1)$ again to transport the discs from the help stack to the third rod. The figure illustrates the case $n = 4$.

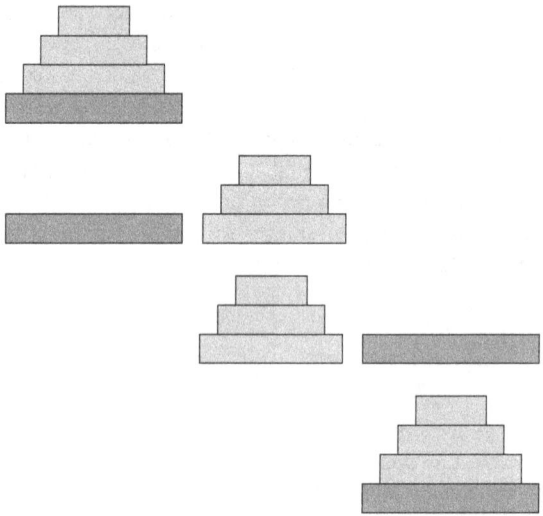

Recursive solution in the case $n = 4$.

In the solution for n discs, the solution for $n - 1$ discs is applied twice and the largest disc is moved exactly once. Therefore, for a complete solution,

$$1 + 2(1 + 2(1 + ...)...) = 1 + 2 + 4 + 8 + \cdots + 2^{n-1} = 2^n - 1$$

discs must be moved. It can be shown that it does not work with fewer moves.

Dedekind showed with the recursion theorem that the natural numbers are unique up to isomorphism. However, he used quantifiers over all subsets of the natural numbers, so from a modern point of view, he based his proofs on second-order predicate logic. Thoralf Skolem later showed that in first-order predicate logic, non-standard models of the natural numbers exist that have different properties than we usually expect. Skolem's exclusive use of first-order logic became a prevailing standard after some time.

Dedekind's approach can be seen as the beginning of structuralism, a mathematical-philosophical view. He used the concept of mapping in a modern sense and thus anticipated the emergence of category theory in the 20th century, which emphasised the structural properties of mathematical objects more than the objects themselves, which are interchangeable.[150]

Recursive calculations and proofs with complete induction were already known before Dedekind, for example by Franziskus Maurolicus, Blaise Pascal and Jakob Bernoulli. In England, Lady Ada Lovelace developed a program for recursive calculations of Bernoulli numbers on the unfinished mechanical calculator planned by Charles Babbage since 1822, which he called the Analytical Engine. Herbert Grassmann, Charles S. Peirce and Giuseppe Peano also considered the axioms of natural numbers, without using the recursion theorem in the way Dedekind did. Peano, like Russell, extensively developed the language of logic from Frege and refined the axioms of natural numbers based on the preliminary work of Dedekind.

Between 1900 and 1930, recursion theory was slowly developed further and a theory of primitive recursive functions emerged, particularly in the works of David Hilbert, Thoralf Skolem and Rózsa Péter. Primitive recursive functions form the smallest class of everywhere defined functions

$$f : \mathbb{N}^n \longrightarrow \mathbb{N}$$

in several variables, which contains constant functions and projection mappings and is closed under the recursion scheme

$$f(0, y) = g(y)$$
$$f(x + 1, y) = h(f(x, y), x, y),$$

where g and h are also primitive recursive functions and the variable x denotes one of the arguments of f and y the rest. In addition to the factorial function, many other elementary functions are primitive recursive. Surprisingly, the function f with

$$f(n) = p_n \text{ (the } n\text{-th prime number)}$$

is also primitive recursive.

The example of recursion theory and the beginnings of the proof-theoretic work of Hilbert and Ackermann clearly show how closely proof theory and computability theory are related. Wilhelm Ackermann and Rózsa Péter dealt with functions that are computable, but not primitive recursive. This includes the Ackermann–Péter function $A(m, n)$, which in Péter's representation from 1935 is given by

$$A(0, n) = n + 1$$
$$A(m + 1, 0) = A(m, 1)$$
$$A(m + 1, n + 1) = A(m, A(m + 1, n)).$$

This function is not primitive recursive, because its growth is stronger than that of any primitive recursive function. In principle, all computable functions can be classified by ordinal numbers using the Löb–Wainer hierarchy. The discovery of rapidly growing functions can be traced back to Hardy. Gödel used this type of characterisation in his famous system T to study the consistency of Dedekind–Peano arithmetic.[151]

The Theory of Computability

The first machine-like models of computability were invented around 1910 by Axel Thue and are now called Thue systems. His standpoint was that every form of computation or proof arises as a sequence of term substitutions. Thue's ideas had a tremendous influence in linguistics, especially through the works of Noam Chomsky and Richard Montague.[152]

After Kurt Gödel and Jacques Herbrand had dealt with a generalisation of primitive recursive functions in a brief correspondence in 1931, which was later improved, it was not until the annus mirabilis 1936 that the concept of computability was clarified with the involvement of several people. The most famous among them is the concept of the Turing machine and thus the term Turing computability by Alan Turing.[153]

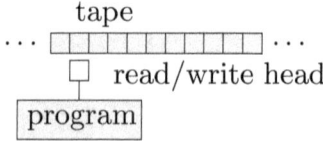

Blueprint of a Turing machine.

The figure outlines the functioning of a Turing machine. The unit labelled as program contains a deterministic function that describes how to proceed from the states and the labels of the tape with an alphabet, i.e., how the read/write head moves and how the tape is newly labelled.

Turing noted that universal Turing machines exist that can simulate any other Turing machine. This is similar to a compiler that makes an executable file from program code. The theoretical concept of Turing machines is therefore very similar to a programmable computer. However, the history of computer development was lengthy. After Leibniz's calculating machine and Babbage's rather unsuccessful attempt to develop his programmable Analytical Engine, the Z3 computer by Konrad Zuse from 1941 and the American computer Mark I from 1944 were the first functioning mechanical freely programmable computers that were universal Turing machines. The Eniac computer from 1946 at the University of Pennsylvania was the first electronic computer of this kind. Today, computers in their internal structure are mostly realised by the von Neumann architecture, which was conceived by John von Neumann and his team during their work on the Eniac.

Emil Post, who was familiar with Thue's work, invented a mathematical theory of automata that resembles Thue systems and Turing machines. Alonzo Church founded the λ-calculus—a third alternative computability theory and today the basis of functional programming languages—and invented the so-called simple theory of types.

Finally, Stephen C. Kleene defined recursive functions, also known as μ-recursive functions, as an extension of the primitive recursive functions and demonstrated the equivalence of these four definitions.[154] For this, he used the μ-operator, which for each partially defined recursive function

$$f(t, x_1, \ldots, x_n) \colon \mathbb{N}^{n+1} \longrightarrow \mathbb{N}$$

forms a new—possibly only partially defined—function μf with

$$\mu f(x_1, \ldots, x_n) \colon \mathbb{N}^n \longrightarrow \mathbb{N},$$

where the value

$$\mu f(x_1, \ldots, x_n) = k,$$

if there is a smallest natural number k such that

$$f(k, x_1, \ldots, x_n) = 0$$

and $f(t, x_1, \ldots, x_n)$ is defined for all $0 \leq t \leq k$. If such a k does not exist, $\mu f(x_1, \ldots, x_n)$ is undefined. The μ-operator can be implemented in programming languages using while-loops by counting up. The class of μ-recursive functions is generated from the primitive recursive functions together with the μ-operator and forms a class that includes partially defined functions. For example, the Ackermann–Péter function can be expressed through the μ-operator.

The μ-operator is by definition a kind of search operator that seeks a root of f in a variable, if one exists. If this is not the case, the resulting function μf is undefined. For example, integer solutions of diophantine equations such as the Catalan equation

$$x^2 - y^3 = 1$$

can be calculated by nested application of the μ-operator—or equivalently using while-loops. In this example, it is best to search on all lines of the form $x + y = n$, where n runs through all natural numbers.[155] Preda Mihăilescu proved the Catalan conjecture in 2002, i.e., this equation only has the solution $(x, y) = (3, 2)$ in natural numbers $x, y \geq 1$.

The Church–Turing thesis, as originally formulated by Alonzo Church and others, states that all effectively computable functions are always recursive. This thesis has been confirmed many times, but it is very informal in terms of the concept of computability and its physical realisation and is therefore considered unprovable. A refutation of the Church–Turing thesis would only be possible through stronger physically realisable computability models, which do not exist to this day.

Artificial neural networks with rational (or computable) weights, as well as the still largely fictional quantum computers, are fascinating computability models. Both fulfil the Church–Turing thesis, as they can in principle be simulated by Turing machines. Non-deterministic Turing machines represent another computability model, which is predominantly of a theoretical nature. Mathematically, this is achieved by replacing the deterministic function in the program of a Turing machine with a relation, i.e., a multi-valued function.

Algorithms are classified in various ways according to objectives or properties.[156] An interesting class are the stochastic algorithms, also called probabilistic algorithms. Effective stochastic algorithms are the Monte Carlo algorithms. An example of this is Monte Carlo integration, where high-dimensional integrals are evaluated by randomly selecting a sufficient number of random support points. Stochastic algorithms are often more efficient than deterministic algorithms. In the case of prime number tests, the Agrawal–Kayal–Saxena prime number test is a deterministic algorithm that provably runs in polynomial time. However, probabilistic prime number tests, such as the Miller–Rabin test or the Solovay–Strassen test, are often much faster in practice. This is no coincidence, as the Miller–Rabin test has a deterministic variant that runs in polynomial time under the assumption of a generalisation of the Riemann hypothesis, because the set of necessary samples can be limited.

A special class of algorithms are the quantum mechanical algorithms. These include the Shor algorithm, which was found by Peter Shor in 1994 for the factorisation of natural numbers, as well as the Deutsch–Jozsa algorithm and Grover's search algorithm. They could be effectively implemented on freely programmable universal quantum computers as soon as they can process enough qubits. Quantum annealers and other commercial quantum computers that are not universal cannot usually execute such algorithms.

Complexity Theory

Even if an algorithm exists for a given problem, it matters greatly how quickly the algorithm works and how much storage space it requires. To increase the performance of algorithms, parallelisation is used, which arranges program parts so that they can be executed simultaneously. However, not every algorithm is suitable for this.

The effort required by an algorithm is considered in complexity theory, which is a separate field of research with many open questions.[157] Computable functions and algorithms that calculate these can be divided into complexity classes depending on the growth or degree of nesting of the functions and the computational and storage effort generated. An important classification of functions is provided by the Grzegorczyk hierarchy. One of the lowest levels of this are the so-called elementarily computable functions, which are sufficient for surprisingly many situations.[158]

A completely different classification is given by the runtime of algorithms. The complexity class **P** is the class of polynomial time algorithms. These are defined by the fact that the runtime (and thus the memory requirement) is polynomial in the effort of the input data.[159] It should be noted that the input of a natural number n into a computable function (or a Turing machine) means an effort of $O(\log(n))$ input data, because the number of digits of n is proportional to $\log(n)$.[160] Algorithms in the class **P** thus have the effort $O(\log^k(n))$ for a natural number k.

Another famous complexity class is the larger class **NP** of computable functions that can be computed by non-deterministic Turing machines with polynomial runtime. One of the Millennium problems of the Clay Foundation asks whether

$$\mathbf{P} = \mathbf{NP}$$

holds or not. Remarkable about this question is that there are problems that are in **NP** and are **NP**-complete. If such a problem is in **P**, then **P** = **NP** already holds. The Millennium problem therefore only needs to be tested on one of them. These include the SAT problem, the travelling salesman problem, and the clique problem.

The SAT problem, also known as the satisfiability problem of propositional logic, consists in deciding for each formula of propositional logic whether there are truth values (true or false) for all variables occurring in it, so that the whole formula becomes true after substituting these values. Obviously, any solution to this problem is verifiable in polynomial time. On the other hand, the number of truth values to be substituted for n variables is of the order of 2^n and thus exponential in n. Stephen Cook and Leonid Levin independently showed that the SAT problem is **NP**-complete.[161]

Quantum computers are potentially much more powerful than classical digital computers, as the Shor algorithm shows. Their polynomial time complexity class is denoted by **BQP**. It lies in the class **PSPACE** of computable functions with polynomial memory effort. In recent years, quantum computers with more than 50 qubits have been built, which can execute specially adapted algorithms faster than classical computers. Presumably, algorithmic problems of a more general nature always require exponentially greater effort with classical computers. This currently very popular assumption refines the Church–Turing thesis and is called the supremacy of quantum computers. Currently, the technology is still far from such expectations.

Undecidable Problems

Surprisingly, there are mathematical problems that cannot be solved algorithmically and are called undecidable. They are based on the existence of undecidable subsets S of the natural numbers. Such sets are recursively enumerable, i.e., either empty or the set of values of a computable function $f : \mathbb{N} \to \mathbb{N}$, but their complement $\mathbb{N} \setminus S$ in $\mathbb{N}$ is not recursively enumerable. In particular, the characteristic function of S is not computable.

Hilbert and Ackermann formulated the decision problem in 1928. It asks whether for every statement in a formal language over first-order predicate logic, it can be decided whether it is provable or not. Alan Turing and Alonzo Church[162] independently showed in 1936 that the decision problem is algorithmically unsolvable and thus undecidable, by showing that the subset S of provable statements is undecidable. The proof is nowadays carried out by reduction to the halting problem for Turing machines. This result relativised the ideas of Leibniz on the decidability of all scientific questions.

The halting problem asks whether there is an algorithm that decides for a given Gödel numbering T_n of Turing machines, which compute partially defined functions $f_n : \mathbb{N} \longrightarrow \mathbb{N}$, whether the function f_n is defined at the argument n. Today's proofs for the undecidability of the halting problem use a variant of Cantor's diagonal argument. In these, assuming that the halting problem is decidable, the total computable function h with

$$h(k) = \begin{cases} f_k(k) + 1 & \text{if } f_k(k) \text{ is defined} \\ 0 & \text{otherwise} \end{cases}$$

is considered. There is then a number n, such that h is computed by the Turing machine T_n corresponding to the function f_n. This yields a contradiction because essentially

$$h(n) = f_n(n) \text{ and } h(n) = f_n(n) + 1$$

would simultaneously hold.

There are numerous other significant problems in mathematics for which no general solution algorithm exists. These include the 10th Hilbert problem and the word problem for semigroups and groups. Somewhat more generally, Rice's theorem states that any non-trivial semantic property of Turing machines is undecidable.[163]

The 10th Hilbert problem asks for an algorithm that, for every integer diophantine equation

$$F(x_1, ..., x_n) = 0$$

decides whether the equation has an integer solution or not. After preliminary work by Martin Davis, Julia Robinson and Hilary Putnam, Yuri Matiyasevich showed in 1970 that the 10th problem is undecidable, i.e., there is no such algorithm. The obvious search algorithm for finding solutions is not effective, because there is no a priori estimate for the size of possible solutions as a termination condition.

The negative solution of the 10th Hilbert problem follows from the existence of undecidable sets and a theorem by Davis, Matiyasevich, Robinson and Putnam, which states that every recursively enumerable set is diophantine and therefore given by the projection of a zero set of a diophantine equation.

Davis, Putnam and Robinson first proved that certain exponential equations, such as the zero set

$$x^y = z$$

in the (x, y, z)-space, are diophantine, although this does not seem so at first glance. Julia Robinson had the idea to use the classical Pell's equation

$$x^2 - dy^2 = \pm 1$$

in the case of $d = a^2 - 1$, whose infinitely many solutions have suitable growth behaviour. The proof by Matiyasevich proceeded somewhat differently, as he used an equation of the form

$$x = F_{2y}$$

in the (x, y)-space, where F_m is the sequence of Fibonacci numbers

$$0, 1, 1, 2, 3, 5, 8, 13, 21, \ldots$$

defined by $F_0 = 0$, $F_1 = 1$ and $F_{m+2} = F_m + F_{m+1}$. Modern proofs usually use Pell's equation as Julia Robinson did.[164]

The word problem for finitely presented groups or semigroups

$$G = \{a_1, \ldots, a_s \mid r_1, \ldots, r_t\},$$

which are given by finitely many generating elements $a_1, \ldots, a_s$ and relations[165] $r_1, \ldots, r_t$, consists in deciding whether an arbitrary word w in G is equal to a given word w_0:

$$w = w_0.$$

So an algorithm is sought that finds finitely many relations, so that the word w goes over into the word w_0 after applying these relations as term substitutions.

The word problem was first formulated by Axel Thue in his already mentioned works on computability and he was aware of the difficulty of this problem. Max Dehn, who had already solved the 3rd Hilbert problem in 1903, formulated the word problem independently from Thue. The presentation of semigroups and groups is closely related to the computability model of Thue systems because the relations r_j can be understood as term substitutions.

The proof for the undecidability of the word problem for semigroups was independently provided by Emil Post and Andrey Markov in 1947. Indeed, the connection with computation models can be used to demonstrate undecidability. Pyotr S. Novikov and William Boone solved the word problem for groups a few years later.

There are important classes of semigroups and groups for which the word problem is solvable. On the other hand, there are specific finitely presented groups for which the problem has no solution. A handy example was found by Gregory Tseytin. For the semigroup

$$G = \langle a, b, c, d, e \mid ac = ca, ad = da, bc = cb, bd = db, ce = eca,$$

$$de = edb, cdca = cdcae, caaa = aaa, daaa = aaa \rangle$$

the problem is undecidable, whether an arbitrary word matches $w = aaa$.[166]

Artificial Intelligence and Hypercomputing

We have dealt with the Platonic world of ideas and its connection with abstract concepts and truth. All this thinking takes place in our heads. It is an interesting and still unsolved question whether human intelligence is superior to a computer or not. Gödel was convinced that this is the case. His argument used ideas around his own incompleteness theorem and is often referred to in slight variations as the Gödel–Lucas–Penrose argument. John Searle's thought experiments go in the same direction.

The answer to this question was not at all obvious in Alan Turing's time. He proposed a related test in an article,[167] which is now called the Turing test. It essentially involves whether a human can be distinguished from a computer through a question-and-answer game. The slowness of the brain and its error-proneness make it clearly inferior to today's computers in many respects. On the other hand, the test neglects some crucial aspects of human intelligence, such as our consciousness, the presence of which may distinguish us from a machine. The Turing test in its historical version is no longer the right question for various reasons.

In recent years, the field of artificial intelligence has come to the fore. A prophet in this field was the Leibniz-inspired mathematician Norbert Wiener, who developed crucial ideas for it as early as the 1950s and at the same time directed a critical, forward-looking view of the future.[168] After many decades of investigating promising approaches, there have only been decisive breakthroughs in implementation in recent years. The currently dominant technology is primarily based on the training of various forms of recurrent artificial neural networks using large amounts of data as input. In many areas, such as image recognition, this method has significantly surpassed any other approach. It is known that recurrent artificial neural networks can simulate any Turing machines.

Recurrent neural networks are based—similar to the multilayer perceptron—on systems of equations

$$v.(n) = \sigma_V \left(W_E \, e.(n) + W_V v.(n-1) + b_V \right)$$
$$a.(n) = \sigma_A \left(W_A \, v.(n) + b_A \right),$$

where $n = 0, 1, 2, \ldots$ is a discrete time parameter, σ_V and σ_A are activation functions, b_V and b_A are biases, and W_E, W_V and W_A are transition operators with weights that operate on the input vectors $e.(n)$ and the hidden variables $v.(n)$ to finally yield the output $a.(n)$. The training of recurrent neural networks is often only performed on the weights of W_A for efficiency reasons.

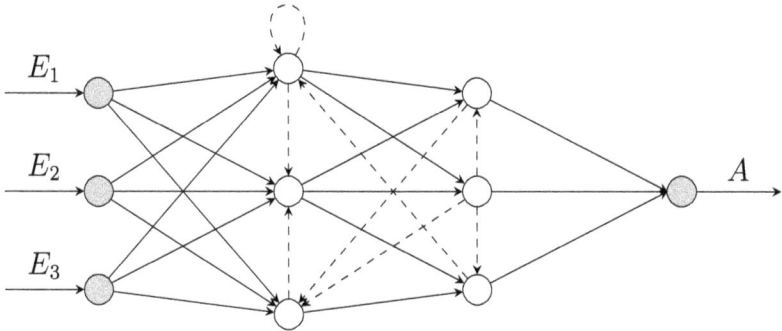

Recurrent neural network (feedback dashed).

It is an open question whether there are other physically realisable computability models, besides quantum computers or the different variants of artificial neural networks, that yield strictly more computing power than classical computers or can even compute functions that are not Turing computable. Such expectations are referred to as hypercomputing. Possible architectures for this are generalisations of recurrent neural networks which contain analogue physical systems as a network structure in the hidden layers, possibly with quantum mechanical components (also known as reservoir computing). As with many other approaches, in addition to the question of hypercomputing, a significant reduction in energy consumption is also sought.

Deductive Systems and Incompleteness

6

In antiquity, scientific discourse developed rapidly. Arithmetic and geometric the-orems were precisely derived under the assumption of axioms. Euclid's influential book "Elements" is an expression of this culture. From today's perspective, every mathematical proof is based on a syntactic calculus, which we refer to as a deductive system. This concept includes a formal language and logical inference rules. After further developments of Aristotelian logic by Llull, Leibniz, Bolzano and others, it was only Frege in his "Begriffsschrift" who introduced a deductive system of today's kind. At the same time, recursion theory and the axiomatics of arithmetic were devel-oped by Dedekind. Shortly afterwards, Peano laid the foundations of today's notation in logic.[169]

After the turn of the 20th century, further perspectives emerged. Axel Thue designed his Thue systems, which represent elementary operations on trees. Hilbert worked together with Wilhelm Ackermann and Paul Bernays on the mathematisa-tion of proofs in axiomatic theories and developed the Hilbert calculus. Movements such as Hilbert's formalism of axiomatic theories, intuitionism and constructivism emerged.[170] These ideas led to the emergence of the field of proof theory, in which one can talk about mathematical proofs in the form of a metamathematics. A high-light of these developments are the two incompleteness theorems by Gödel. They form a fundamental obstacle for Hilbert's dream of a proof of the consistency of mathematical theories, which only appears partially surmountable through the use of transfinite numbers or other axioms.

Formal Languages and Deductive Systems

A deductive system is a calculus in which proofs can be conducted. The basis of every deductive system is a formal language along with some axioms and inference rules. A proof in a deductive system uses the axioms and inference rules in a suitable order until the desired result is achieved.[171]

Let's first turn to formal languages. The classical, two-valued propositional logic is one of the simplest formal languages with the special characters

$$\neg, \wedge, \vee, \Rightarrow,$$

which we have already learned. Through the relationships

$$A \vee B = \neg(\neg A \wedge \neg B), \quad A \Rightarrow B = \neg(A \wedge \neg B)$$

it is sufficient to only consider the logical symbols $\wedge$ and $\neg$. Propositional logic has some variants, most of which are unknown to most people. There are multi-valued logics, modal logic and intuitionistic logic.

First-order predicate logic extends the propositional logic and consists of a reservoir of symbols for constants, free and bound variables, function and relation symbols (predicates), as well as the additional symbols

$$\forall \text{ and } \exists,$$

which we have already learned. While function symbols can take all permissible values within the language, predicates provide the truth values true or false. A vivid example of a unary predicate is a property of the form

$$P(x): x \text{ is red.}$$

The equality sign $=$ is usually part of every formal language as a binary relation or as a binary predicate. In a formal language L, there are usually other non-logical symbols. Depending on the language, these can be constants like 0 and 1, unary function symbols like S (successor mapping), minus sign, inverse symbol or binary function symbols like $+$, $\cdot$, $\circ$ or relation symbols like $<$ and $\in$.

From the constants, the variables and the function or relation symbols of a formal language L, terms can be formed. All constants and free variables are terms and the values of functions $f(t_1, \ldots, t_n)$, into which terms $t_1, \ldots, t_n$ have been inserted, are again terms.

From terms, formulas can be formed, with atomic formulas arising by inserting terms into predicates $P(t_1, \ldots, t_n)$. At this point, the equality sign $=$ can occur. General formulas arise from atomic formulas by using the logical symbols, i.e., if P and Q are formulas, then $\neg P$, $P \wedge Q$, $P \vee Q$ and $P \Rightarrow Q$ are formulas. If $P(x)$ is a formula containing a free variable x, then

$$\exists x \; P(x) \text{ and } \forall x \; P(x)$$

are again formulas. After all, a sentence is a formula which is closed, i.e., without free variables. Such expressions can be viewed as statements (or propositions, theorems) which have a truth value.

Calculi include, in addition to a formal language, additional axioms and inference rules. In all deductive systems, substitution is one of the inference rules. The notation

$P[\frac{t}{x}]$ is common when the expression t is substituted for x in the formula $P(x)$. When performing substitutions in sequence, the equality $P[\frac{t}{x}][\frac{s}{t}] = P[\frac{s}{x}]$ is usually required. The calculi of Frege, Russell and the Hilbert calculus possess numerous logical axioms and have only the modus ponens

$$\frac{P \quad P \Rightarrow Q}{Q}$$

as a new inference rule. The Hilbert calculus and preceding calculi of Frege and others correspond little to the usual methods of mathematical reasoning and the logical axioms are unnatural in some versions. Therefore, from 1926, Jan Łukasiewicz, Stanisław Jaśkowski and Gerhard Gentzen introduced new calculi, of which we will use the calculus of natural deduction.[172] This calculus uses judgements of the form

$$\Gamma \vdash A,$$

where Γ denotes a finite number of hypotheses and A is a single formula. The symbol $\vdash$ goes back to the Fregean judgement stroke and expresses the derivability of A based on the assumption Γ in the calculus. It is the metalinguistic version of implication and must be distinguished from the object language symbol $\Rightarrow$. If Γ is empty, then $\vdash A$ denotes a formula A that is derivable without hypotheses, i.e., an axiom or a theorem.

In propositional logic, there are two tautological inference rules

$$\frac{}{A \vdash A}, \quad \frac{\Gamma \vdash A}{\Gamma' \vdash A} \text{ for } \Gamma' \supseteq \Gamma$$

and two symmetry rules

$$\frac{\Gamma \vdash (A \wedge B)}{\Gamma \vdash (B \wedge A)}, \quad \frac{\Gamma \vdash (A \vee B)}{\Gamma \vdash (B \vee A)}.$$

In addition, there are four introduction rules each for $\wedge, \vee, \neg, \Rightarrow$

$$\frac{\Gamma \vdash A \quad \Gamma \vdash B}{\Gamma \vdash A \wedge B}, \quad \frac{\Gamma \vdash A}{\Gamma \vdash A \vee B}, \quad \frac{\Gamma, A \vdash B \quad \Gamma, A \vdash \neg B}{\Gamma \vdash \neg A}, \quad \frac{\Gamma, A \vdash B}{\Gamma \vdash (A \Rightarrow B)}$$

and the corresponding four elimination rules

$$\frac{\Gamma \vdash A \wedge B}{\Gamma \vdash A}, \quad \frac{\Gamma \vdash A \vee B \quad \Gamma, A \vdash C \quad \Gamma, B \vdash C}{\Gamma \vdash C}, \quad \frac{\Gamma \vdash \neg \neg A}{\Gamma \vdash A}$$

and

$$\frac{\Gamma \vdash A \quad \Gamma \vdash A \Rightarrow B}{\Gamma \vdash B} \text{ (modus ponens)}.$$

One of the most important discoveries of Gentzen is his main theorem of cut elimination with the provable inference rule

$$\frac{\Gamma \vdash A \quad \Delta, A \vdash B}{\Gamma, \Delta \vdash B},$$

which allows the formula A to be eliminated. This rule in turn implies the two symmetry rules. Gentzen used the cut elimination rule to show the consistency, i.e., the contradiction-freeness, of Dedekind–Peano arithmetic.

The ability to find and elegantly conduct correct mathematical proofs is a great art. George Pólya wrote a wonderful book about it in 1945 titled "How to solve it".[173] An unspectacular proof of a simple theorem like

$$\vdash (A \wedge B) \wedge C \Rightarrow A \wedge (B \wedge C),$$

which—due to the rules applicable to $\Rightarrow$—is equivalent to

$$(A \wedge B) \wedge C \vdash A \wedge (B \wedge C),$$

consists of the derivation tree shown. The proof uses the abbreviation $\Gamma = (A \wedge B) \wedge C$ and—starting from the tautology $\Gamma \vdash \Gamma$—uses conclusion rules at the solid lines.

$$
\cfrac{
\cfrac{\Gamma \vdash (A \wedge B) \wedge C}{\Gamma \vdash A \wedge B}
\qquad
\cfrac{
\cfrac{\cfrac{\Gamma \vdash (A \wedge B) \wedge C}{\Gamma \vdash A \wedge B}}{\cfrac{\Gamma \vdash B \wedge A}{\Gamma \vdash B}}
\qquad
\cfrac{\cfrac{\Gamma \vdash (A \wedge B) \wedge C}{\Gamma \vdash C \wedge (A \wedge B)}}{\Gamma \vdash C}
}{\Gamma \vdash B \wedge C}
}{\Gamma \vdash A \wedge (B \wedge C)}
$$

Derivation tree of $(A \wedge B) \wedge C \vdash A \wedge (B \wedge C)$.

In predicate logic, additional introduction rules for $\exists$, $\forall$ and $=$ are needed. These are

$$\frac{\Gamma \vdash P[\frac{t}{x}]}{\Gamma \vdash \exists x\, P(x)}, \quad \frac{\Gamma \vdash P[\frac{t}{x}]}{\Gamma \vdash \forall x\, P(x)} \ (t \text{ free}), \quad \frac{}{\Gamma \vdash P = P}.$$

The corresponding elimination rules are

$$\frac{\Gamma \vdash \exists x\, P(x) \quad P[\frac{t}{x}] \vdash Q}{\Gamma \vdash Q} \ (t \text{ free}), \quad \frac{\Gamma \vdash \forall x\, P(x)}{\Gamma \vdash P[\frac{t}{x}]}, \quad \frac{\Gamma \vdash s = t \quad \Gamma \vdash P[\frac{t}{x}]}{\Gamma \vdash P[\frac{s}{x}]}.$$

The adjective free here means that the variable t appears independently of the other occurring variables and constants and can take any values.

The Dedekind–Peano Arithmetic

We now want to deal with the formal language of Dedekind–Peano arithmetic. This language is used to describe the foundations of arithmetic, i.e., the theory of natural numbers $\mathbb{N}$ and their additive and multiplicative properties.

The language of Dedekind–Peano arithmetic in addition to the logical symbols $\wedge$, $\vee$, $\neg$, $\Rightarrow$, $\forall$, $\exists$, consists of the symbols 0 (zero), S (successor function), $+$ (addition), $\cdot$ (multiplication) and $=$ (equality). A typical formula in it is given by the 4-squares theorem

$$\forall n \, \exists a \, \exists b \, \exists c \, \exists d \quad n = a^2 + b^2 + c^2 + d^2,$$

which states that every natural number n can be represented as the sum of four squares. In the calculus of this language, the usual conventions such as "dot before line", bracket symbols and other abbreviations are allowed. In Dedekind–Peano arithmetic there are the following axioms:

$$\forall n \quad S(n) \neq 0 \tag{6.1}$$
$$\forall m \forall n \quad S(m) = S(n) \Longleftrightarrow m = n \tag{6.2}$$
$$\forall n \quad n + 0 = n = 0 + n \tag{6.3}$$
$$\forall m \forall n \quad m + S(n) = S(m + n) = m + S(n) \tag{6.4}$$
$$\forall n \quad n \cdot 0 = 0 = 0 \cdot n \tag{6.5}$$
$$\forall m \forall n \quad m \cdot S(n) = m \cdot n + m \tag{6.6}$$
$$P(0) \wedge \forall n \, (P(n) \Rightarrow P(S(n))) \Rightarrow \forall n \, P(n). \tag{6.7}$$

The successor function S is noted as

$$S(n) = n + 1$$

when addition has already been defined by recursion. So it holds:

$$1 = S(0), \; 2 = S(S(0)), 3 = S(S(S(0))), \; \ldots$$

The most important axiom is the last axiom (7) of complete induction. For every property P of natural numbers which can be formulated in the formal language of Dedekind–Peano arithmetic it expresses the following statement:

If $P(0)$ holds and if $P(n)$ always implies $P(S(n))$, then P is fulfilled for every natural number.

This statement can be interpreted as a rule of inference

$$\frac{P(0) \quad \forall n \, (P(n) \Longrightarrow P(S(n)))}{\forall n \, P(n)}.$$

In Dedekind–Peano arithmetic the order relation $<$ is also of great importance. This relation can be added to the language.

Gödel's First Incompleteness Theorem

Kurt Gödel spent a lot of his life intensely studying Leibniz and the idea of the Scientia generalis. He became so engrossed in the reading that he occasionally believed that the editions of Leibniz's works were deliberately withholding certain content when publishing.

Gödel revolutionised mathematical logic around 1930 already in his dissertation supervised by Hans Hahn with the proof of the completeness theorem. He succeeded shortly afterwards, in his work "On formally undecidable propositions of Principia mathematica" from 1931, to form the self-referential Gödel sentence Q within the Dedekind–Peano arithmetic which—colloquially speaking—is of the form

> I am not provable.

He wrote in the introduction to this work, which summarises the entire proof, that he was inspired by the paradox of Epimenides for his proof.[174]

In it, Gödel used the method of Gödelisation, to number sequences of symbols in formulas and proofs of Dedekind–Peano arithmetic with the help of elementary number theory. There are different ways to define Gödel numbers for sequences of symbols. One variant uses the sequence of prime numbers

$$p_1 = 2, \quad p_2 = 3, \quad p_3 = 5, \ldots$$

and assigns a natural number to each symbol, for example

$$A \to 1, B \to 2, C \to 3, \ldots, H \to 8, \ldots$$

These numbers are used as exponents of the prime numbers. Thus, a sequence of symbols like BACH is given the Gödel number

$$2^2 \cdot 3^1 \cdot 5^3 \cdot 7^8 = 8647201500.$$

Similar numberings of expressions in formal languages already occur with Leibniz. The Gödel number of a formula P is denoted by $\ulcorner P \urcorner$. The unique prime factor decomposition of natural numbers makes it possible to reconstruct the sequence of symbols BACH from the Gödel number 8647201500.

Since proofs of statements of arithmetic are finite sequences of formulas, Gödel had to show that the proof operations in the deductive system of Dedekind–Peano arithmetic correspond to primitive recursive functions of the Gödel numbers. This property allowed him to form the Gödel sentence Q that corresponds to the colloquial statement above in a syntactic way.

How does Gödel's proof work in detail? The proof of the theorem uses class signs (German Klassenzeichen), i.e., formulas of the form $\varphi(x)$ which depend on a free variable. These can also be assigned Gödel numbers $\ulcorner \varphi(x) \urcorner$. It is possible to number the class signs with an index k so that $\varphi_k(x)$ receives the Gödel number k. Not all natural numbers appear as an index. Gödel showed that the mapping diag, which sends the number k to the Gödel number $\ulcorner \varphi_k(k) \urcorner$ of $\varphi_k(k)$, is primitive recursive. This part of the proof is reminiscent of Cantor's diagonal argument. With these preconditions, he proved that the two-place provability relation

$$\text{Proof}(x, y) = \begin{cases} x \text{ encodes a proof} \\ \text{of the formula } \varphi_y(y) \end{cases}$$

is primitive recursive and can be formulated syntactically in the Dedekind-Peano arithmetic. Thus, using diag, $\neg$ and Proof, the class sign

$$\varphi(y) = \forall x \, \neg\text{Proof}(x, y) = \begin{cases} \text{For all } x \text{ it is not the case that } x \\ \text{encodes a proof of the formula } \varphi_y(y) \end{cases}$$

can be defined. Therefore, there is an n with $\varphi = \varphi_n$ and the sought-after Gödel sentence can be written as

$$Q = \varphi_n(n).$$

It now easily follows that Q is provable if and only if Q is not provable and vice versa. This results in a contradiction as in the paradox of Epimenides.[175]

With this, Gödel had proven the first incompleteness theorem, by asserting the existence of unprovable and irrefutable sentences Q in any formal languages that can formulate the elementary features of Dedekind–Peano arithmetic. A shorter proof of this theorem was found by Raymond Smullyan. He constructed a minimal logical language in which the self-referential Gödel sentence Q can be formulated.[176]

In modern proofs, Gödel's trick is usually explained with the diagonal lemma which was implicitly contained in Gödel's proof. It was first explained by Gödel in a lecture at the Institute for Advanced Study in Princeton in 1934 and he said that it was based on an idea by Rudolf Carnap.[177] The diagonal lemma states that every class sign $P(x)$ has a sentence Q as a fixed point, for which

$$Q \Longleftrightarrow P(\ulcorner Q \urcorner)$$

is provable. As class sign $P(x)$, with the help of the unary provability predicate $\text{Proof}(y) = \exists x \text{Proof}(x, y)$, the predicate

$$P = \neg\text{Proof}$$

of non-provability can be chosen, i.e., $P(n)$ is true if and only if n is not the Gödel number of a provable formula Q. From the diagonal lemma now results the famous formula

$$Q \Longleftrightarrow \neg\text{Proof}(\ulcorner Q \urcorner)$$

from which—as in the original proof by Gödel—it follows that Q is neither provable nor refutable.

Per Martin–Löf has connected the Gödelian concept of incompleteness with Kantian synthetic judgements.[178] The incompleteness corresponds to the information hidden in a synthetic judgement which cannot be obtained analytically.

Gödel's Second Incompleteness Theorem

The second incompleteness theorem states that the proof of consistency, i.e., the contradiction-freeness, of mathematical theories that contain Dedekind–Peano arithmetic, is not possible within the same. The contradiction-freeness can be expressed by the syntactic arithmetic formula Con which corresponds to the statement

$$\neg\text{Proof}(\ulcorner 0 = 1 \urcorner).$$

This theorem has a surprising history. A proof for it was not included in the submitted manuscript and was only announced at the end of Gödel's 1931 publication in a somewhat cryptic remark as a corollary for a second part that never appeared. Possibly, Gödel had this passage inserted shortly before printing because he feared a competing publication by John von Neumann, who wrote him a letter outlining this result on November 20th 1930. In his responses to von Neumann, Gödel presented the argument himself and a proof was later published by Hilbert and Bernays.[179] The proof of the incompleteness theorem only requires the contradiction-freeness Con. Therefore, with additional arguments, the implication

$$\text{Con} \Rightarrow Q$$

can be proven. Since the Gödel sentence Q is unprovable within the theory, this also applies to Con. The consistency of the more general Zermelo–Fraenkel set theory also cannot be derived as a formula, because it would—by virtue of the standard model $\mathbb{N}$—imply the consistency of the subsystem of arithmetic.

This was a spectacular result, as it dampened the hopes of Hilbert's programme to establish the proof theory and thus metamathematics on exact foundations. Neither Hilbert nor Gödel saw the second incompleteness theorem as devastating for Hilbert's programme though. Instead, both suggested independently shortly afterwards to modify the finite standpoint and continue Hilbert's programme. Hilbert wrote in 1934:

> With regard to this goal, I would like to emphasise that the temporarily arisen opinion, that certain recent results by Gödel imply the infeasibility of my proof theory, has proven to be erroneous. That result indeed only shows that for the more advanced proofs of consistency one must exploit the finite standpoint in a sharper way than is necessary when considering elementary formalisms.[180]

Gödel expressed this in 1931 as follows:

> It should be expressly noted that theorem XI (...) is in no contradiction to Hilbert's formalistic standpoint. For this only presupposes the existence of a proof of consistency conducted with

finite means and it would be conceivable that there are finite proofs that cannot be represented in P [Gödel's formal system] (...).[181]

Even many years later, around 1961–1962, Gödel, according to Gerald E. Sacks, occasionally remarked that Hilbert's programme was still open.

The above-mentioned requirements for Con as a syntactic arithmetic formula, which also apply in all non-standard models, are very strong. Therefore, attempts have been made to express consistency either through a different formula with a deviating proof predicate or through an infinite schema of formulas. As a second possibility, Gentzen and Ackermann used transfinite induction up to the ordinal number ε_0. In the even more general case of set theory, consistency can be demonstrated in a third way using Grothendieck universes which are based on additional axioms about inaccessible cardinal numbers. The price for consistency proofs of the last two types are new axioms which again generate incompleteness. In the first case, the concept of consistency is different.[182]

Vladimir Voevodsky expressed doubts about the consistency of mathematics in a remarkable lecture in 2010. He saw type theory and intuitionism as a possible way out. This happened around the same time as Edward Nelson from Princeton—a committed opponent of Platonism and semantics—unsuccessfully tried to refute the consistency of arithmetic. The Hilbertian programme is, all things considered, indeed still partially open and new insights into it are conceivable.

The Gödel–Lucas–Penrose Argument

Most people are probably convinced that human qualities and qualia such as empathy, consciousness, intuition, understanding, feelings and other factors distinguish our thinking from a Turing machine. However, such opinions are far from a scientifically tenable thesis. The opposite idea is that human thinking and behaviour as a whole is equivalent to a universal Turing machine. This position is called the mechanistic thesis.

Such questions were what Alan Turing wanted to decide. He formulated the Turing test, a hypothetical experiment, in which targeted questions are used to determine whether an interlocutor is a human or a computer. John Searle's thought experiment of the Chinese Room has a similar approach.[183]

In Gödel's estate, an unpublished speech manuscript for the Gibbs Lecture of the American Mathematical Society at Christmas 1951 was found, from which it became clear that he believed in the superiority of human thinking. He outlined an argument that used his incompleteness theorem:

Evidently no well-defined system of correct axioms can comprise all objective mathematics, since the proposition which states the consistency of the system is true, but not demonstrable in the system. However, as to subjective mathematics, it is not precluded that there should exist a finite rule producing all its evident axioms ... The assertion, however, that they are all true could at most be known with empirical certainty, on the basis of a sufficient number of instances or by other inductive inferences. If it were so, this would mean that the human mind (in the realm of pure mathematics) is equivalent to a finite machine that, however, is unable to understand completely its own functioning ... So the following disjunctive conclusion is inevitable: Either mathematics is incompletable in this sense, that its evident axioms can never be comprised in a finite rule, that is to say, the human mind (even within the realm of pure mathematics) infinitely surpasses the powers of any finite machine, or else there exist absolutely unsolvable diophantine problems of the type specified.[184]

Gödel derived here an alternative between the opposite of the mechanistic thesis and the existence of certain arithmetic propositions. Moreover, he considered the possibility that the human brain could be equivalent to a Turing machine that does not fully understand its own functioning. At the end of his speech, Gödel tried to substantiate the view of Platonic idealism with similar arguments.

The exact proof argument is not given precisely in Gödel's text. The idea is to consider the deductive system S of Dedekind–Peano arithmetic. Gödel's first incompleteness theorem provides a method to generate a more complete deductive system S' by adjoining the Gödel sentence Q_S of S. Alternatively, the statement Con_S of the consistency of S can be adjoined and a new Gödel sentence $Q_{S'}$ in S' results. The extension S' is called consistency extension. Such consistency extensions can be repeated through infinite transfinite iterations over ordinal numbers.[185] Among all the true theorems in the standard model in the Dedekind–Peano arithmetic there are still those after such iterations that are not provable. There are now two options. Either man can prove all true theorems of the Dedekind–Peano arithmetic. In this case, Gödel's first incompleteness theorem implies that the human brain is infinitely superior to any Turing machine and the mechanistic thesis is false. Otherwise, there is a true arithmetic theorem that is absolutely unsolvable, as Gödel puts it.

John Randolph Lucas tried in 1961 to develop a similar argument in his own way. Roger Penrose also published variants of Gödel's and Lucas' arguments in two books from 1989 onwards. Lucas and Penrose went beyond Gödel and claimed that the mechanistic thesis is always false. We do not want to fully reproduce the exact arguments here, as there are gaps in the proof that were uncovered by the two logicians Martin Davis and Solomon Feferman and could never be eliminated despite many debates. This was merely criticism of the proof itself, as the two aforementioned experts, like Gödel, Lucas and Penrose, were convinced of the superiority of human thought.[186]

The thought processes of Penrose and Lucas are occasionally used in popular science literature in an unscientific way as arguments to refute the mechanistic thesis and to draw all possible conclusions from it. Without proof, however, such claims remain mere speculation.

In this context, Roger Penrose conjectured that the human brain needs quantum mechanical mechanisms to exceed the limit of Turing-computability. At the same time, this could explain the phenomena of consciousness and free will. He argued that the collapse of the wave function in quantum mechanics plays a crucial role in this. However, quantum mechanical effects in the functioning of the brain have never been proven to this day. The free will of humans is at least questionable due to neuroscientific experiments, as human decisions are often preceded by signals in the brain by fractions of a second.

Intuitionism

Proofs of classical logic occasionally use contradiction proofs and thus the law of excluded middle, called tertium non datur in Latin. This method is not universally accepted. Intuitionism is a movement in mathematics coined by Hermann Weyl, Luitzen Egbertus Jan Brouwer and Arend Heyting. It has a related manifestation in constructivism. Both views result from the desire for mathematics to be carried out through a comprehensible thought process of humans. The existence of mathematical objects should be verifiable and the truth of statements provable. The use of the law of excluded middle is therefore rejected and the axiom of choice is considered a tautology.[187] This attitude is historically explainable from the antinomies of set theory at the beginning of the 20th century. The law of excluded middle states

$$\vdash A \vee \neg A$$

for every statement A. This is equivalent to the following equivalent judgements in the calculus of natural deduction, in which A and B are arbitrary:

- $\neg\neg A \vdash A$
- $\vdash ((A \Rightarrow B) \Rightarrow A) \Rightarrow A$ (Peirce's law).

Both inference rules therefore do not apply in intuitionistic mathematics. The reversal

$$A \vdash \neg\neg A$$

applies without additional assumptions and follows from the calculus with the introduction rule for $\neg$ (with $\neg A$ instead of A and $\Gamma = B = A$).

Brouwer was a charismatic person with a varied life. He already dealt with intuitionism in his dissertation. After a creative phase in which he was able to prove deep results of topology, such as the invariance of dimension and his fixed point theorem, he turned to the foundations of mathematics around 1913. The famous foundational dispute between 1921 and 1928 led to a bitter dispute with Hilbert, who had little appreciation for intuitionism. This was followed by the exclusion of Brouwer from the editorial board of the Mathematical Annals. Arend Heyting developed intuitionistic propositional logic and Heyting arithmetic as the intuitionistic counterpart of Dedekind–Peano arithmetic. Hermann Weyl adapted the ideas of intuitionism in his own way and also acted against Hilbert during the foundational dispute.[188]

The ideas behind intuitionism and some metamathematical positions of Brouwer were controversial from the start. His concern was that man as a creative subject constructs the objects of mathematics in his mind and works out the proofs of statements. In particular, he thought a lot about the continuum and wanted to understand it better with the help of so-called choice sequences. Many ideas of Brouwer were influential and have been further developed later. Especially Stephen C. Kleene, George Kreisel, John R. Myhill and Richard E. Vesley have further elaborated intuitionistic analysis in the 1960s.

All logical connections and quantifiers are considered as conditions to be verified in intuitionism. For example, a statement of the form $A \vee B$ is provable if there is a proof of A or a proof of B. All other logical signs can be considered in a similar form. This idea is called Curry–Howard correspondence. Gödel noted that while intuitionism appears as a restriction of classical mathematics, it can also be seen as an extension. To this end, he constructed—with the help of double negation—an interpretation of first-order classical logic in intuitionistic logic using a variant of the following recursive assignment $\varphi \mapsto \varphi^n$:

$$\varphi^n = \neg\neg\varphi, \text{ if } \varphi \text{ is atomic}$$
$$(\varphi \wedge \psi)^n = \varphi^n \wedge \psi^n$$
$$(\varphi \vee \psi)^n = \neg(\neg\varphi^n \wedge \neg\psi^n)$$
$$(\varphi \Rightarrow \psi)^n = (\varphi^n \Rightarrow \psi^n)$$
$$(\neg\varphi)^n = \neg\varphi^n$$
$$(\forall x \, \varphi)^n = \forall x \, \varphi^n$$
$$(\exists x \, \varphi)^n = \neg\forall x \, \neg\varphi^n.$$

Gödel applied this interpretation to arithmetic and showed that every contradiction derivable in Dedekind–Peano arithmetic is also derivable in Heyting arithmetic. Consequently, Heyting arithmetic and Dedekind–Peano arithmetic are equiconsistent.[189] Moreover, Gödel's incompleteness theorems with their consequences also apply in an intuitionistic version. This makes it clear that intuitionism can only alleviate concerns about classical logic to a limited extent.

Constructivism and the Question of Existence

An influential variant of intuitionism is constructivism. The existence of mathematical objects is a crucial concept in both currents. Behind this is the idea that mathematical objects, whose existence is asserted in a theorem, must always be concretely constructed. This implies that proofs by contradiction and the axiom of choice in its classical formulation are not permitted. It is common to regard the axiom of choice as a tautology by interpreting the specification of any number of non-empty sets as the stronger assumption of specifying just as many explicit elements. Since mathematics is an abstract science and the Platonic world of ideas raises doubts, the kind of existence that arises from a constructive approach seems to be more nominalistic and not Platonic in nature. Therefore, it is not clear whether the intuitionistic idea of construction changes anything about the quality of existence.

What exactly construction means is interpreted differently in different variants. An unproblematic requirement for constructability is that concrete objects should be named for mathematical existence claims to generate evidence. A good example of this is the 4-squares theorem, which provides a representation

$$n = a^2 + b^2 + c^2 + d^2$$

for every natural number n. For a given n, the explicit specification of an algorithm that calculates the quadruple (a, b, c, d) depending on n is such evidence. The proof of this theorem can be conducted constructively in this sense.[190]

Hilbert stirred up the movements of intuitionism and constructivism through his attitude towards mathematics, which was particularly manifested in formalism. In a letter dated December 29th 1899 to Frege, he wrote:

> If the arbitrarily set axioms do not contradict each other with all their consequences, then they are true, then the things defined by the axioms exist. This is for me the criterion of truth and existence.[191]

Hilbert experienced strong counter-reactions due to his clear views. Especially Brouwer and Oskar Becker as well as occasionally Weyl fought against Hilbert. In retrospect, it is not necessary to assume that Hilbert meant the word existence in a Platonic or similar sense. He was probably more concerned with being able to work with such objects without contradiction, even if they were not constructively generated. This is a legitimate standpoint, as Hilbert had clearly recognised that the consistency of mathematics is the real problem and not the supposedly missing existence of its objects.

What are the consequences of the constructivist viewpoint? Let's first consider the real numbers. Any $a \in \mathbb{R}$ is constructively given by a Cauchy sequence $(a_n)_{n \in \mathbb{N}}$ that converges to a and fulfils a suitable error estimate. At least from a classical point of view, the trichotomy

$$a < 0 \text{ or } a = 0 \text{ or } a > 0$$

certainly applies. However, for a given Cauchy sequence $(a_n)_{n \in \mathbb{N}}$, it is not constructively decidable which of the three cases applies. This phenomenon is well-known in numerical mathematics. The step function

$$f(x) = \begin{cases} 0 & \text{for } x < 0 \\ 1 & \text{for } x \geq 0 \end{cases}$$

is—in other words—not meaningfully definable in a constructive sense. This implies, for example, that there is no characteristic function $\chi_{\mathbb{Q}}$ of the subset of rational numbers in the real numbers. Brouwer had already seen this and he proved that in the constructive context every total real function is continuous and uniformly continuous on compact intervals. Not least because of such consequences, the intuitionistic and the constructive view have not prevailed widely to this day.

There are mathematical propositions for which no constructive proof is known. An example of this is König's lemma, which was found by Dénes König in 1936.[192] It states that a connected graph, in which only infinitely many edges spring from each node, is exactly infinite, when there is an infinite path in it that avoids itself. In the special case of a finitely branched, infinite tree, we are looking for an infinite path. The proof of this lemma in classical logic is not very difficult. It determines at each step by means of complete induction a new edge, over which infinitely many other nodes can be reached. The law of excluded middle is used to show that such exists.[193]

In the time after Brouwer, the schools of Andrey A. Markov, Nicolai A. Shanin and Errett Bishop further developed constructive mathematics. This resulted in new approaches that can handle large parts of mathematics. The search for those theorems of mathematics that have constructive proofs provides extremely valuable insights. Even though intuitionism and constructivism still play a subordinate role, they are nevertheless significant as they form the foundations of intuitionistic dependent type theory and underlie the internal syntax of categories.[194]

Category Theory

7

Category theory is—alongside set theory and type theory—one of the three possible foundations of mathematics. It incorporates the structural mathematical thinking in an abstract way and is well-suited for the justification of a mathematical semantics. Richard Dedekind has in a certain sense founded structuralism in mathematics. He realised that mathematical objects, such as numbers, can have many isomorphic set realisations. It is only their operations and structural properties that are unique. Dedekind has axiomatically characterised the natural numbers in his book "What are and what should the numbers be?" as chains and proved their uniqueness up to isomorphism with the recursion theorem.[195]

Emmy Noether has significantly contributed to establishing these ideas of Dedekind in modern algebra and in topology, where they were particularly fruitful.[196] In the form of category theory, structuralism was further developed by the topologists Saunders MacLane and Samuel Eilenberg from about 1945 onwards and then found its way into the works of Alexander Grothendieck, Daniel Quillen and others. Especially the development of model and path categories has shown the importance of category theory in homotopy theory and in related algebraic areas, because they conceptualise the role of (weak) equivalences in a conceptual way. An important role in current research is played by higher categories and infinity categories as further developments. From model categories infinity categories can be obtained through localisation.[197]

Categories

A category $\mathcal{C}$ consists of a collection $\mathrm{Ob}(\mathcal{C})$ of objects that share common structural features of a given mathematical concept. Between any two objects A and B of $\mathcal{C}$ there is a collection $\mathrm{Hom}_{\mathcal{C}}(A, B)$ of structure-preserving morphisms, which are represented in the form of arrows

$$A \xrightarrow{\;f\;} B.$$

For each object A in $\mathcal{C}$ there is a natural morphism $\mathrm{id}_A : A \longrightarrow A$, which is called identity.

The collections of objects and morphisms do not necessarily have to be sets in a given set theory, even though this is often required. In such a case, the category $\mathcal{C}$ is called small. We will try as far as possible to consider category theory as an alternative basis for mathematics and to avoid sets when studying categories.

Publications in category theory are usually full of diagrams of arrows. The corresponding morphisms are chained by composition (see diagram).

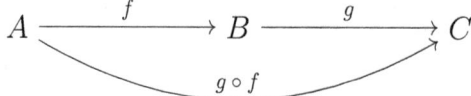

Composition of morphisms.

For this, the associative law applies—similar to multiplication

$$h \circ (g \circ f) = (h \circ g) \circ f.$$

A morphism $f : A \longrightarrow B$ is called an isomorphism if there is an inverse morphism $f^{-1} : B \longrightarrow A$ that fulfils $f^{-1} \circ f = \mathrm{id}_A$ and $f \circ f^{-1} = \mathrm{id}_B$.

Arrows can be chained in different ways. A typical situation is a commutative diagram for which

$$i \circ h = g \circ f$$

applies, i.e., it does not matter which path the square diagram is traversed. This results in a unique new arrow, which is drawn dashed as a diagonal.

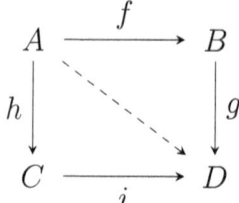

Commutative diagram.

There are many different categories in mathematics. Important examples are the categories **Set** and **Top**. In **Set** the objects are sets and the morphisms are the set-theoretic mappings. In **Top**, the objects are the topological spaces and the morphisms are the continuous mappings. A special category is assigned to each topological space X, namely the category **Off** (X) of open sets in X. The objects therein are the open sets U in X and the morphisms are the inclusions between them. Another important category **Ch** consists of chain complexes of abelian groups. It contains objects such as the singular chain complexes $\mathbb{Z}\text{Sing.}(X)$ of topological spaces.

It is possible, in a given category $\mathcal{C}$, to reverse all arrows. This results in the opposite category $\mathcal{C}^{\text{op}}$.

Groupoids

A very simple category is assigned to each group G. It has only one object $*$ and for each $g \in G$ an arrow, which is also denoted by g (see diagram).

$$
\begin{array}{c}
g \\
\curvearrowright \\
* \rightleftharpoons g^{-1}
\end{array}
$$

$$\text{Category of a group.}$$

This category is a groupoid, i.e., a category, in which every morphism f is an isomorphism. A groupoid is a far-reaching generalisation of a group.

An interesting groupoid, which does not come from a group in this form, is the fundamental groupoid $\Pi_1(X)$ for a topological space X. It is defined as the category, whose objects are the points of X and whose morphisms between a and b are the homotopy classes of paths from a to b. From this fundamental groupoid $\Pi_1(X)$, both the set $\pi_0(X)$ of connected components of X and the fundamental group $\pi_1(X, *)$ for every base point $*$ can be reconstructed.

This groupoid contains the homotopy classes of all paths in X and thus more information than the fundamental group $\pi_1(X, *)$. To go from homotopy classes of paths to all paths, a further step of abstraction is necessary, which will lead us from categories to infinity categories.

Functors

Like all mathematical objects, categories like **Set** or **Top** are not unique. There are variants of **Set** which depend on chosen axioms. For example, the sets in **Set** are often restricted to subsets of a fixed Grothendieck universe U. Some significant theorems of mathematics assert the equivalence of two categories. We therefore need suitable tools to compare categories with each other. A covariant functor between two categories is an arrow

$$F : \mathcal{A} \longrightarrow \mathcal{B}$$

which sends objects A in $\mathcal{A}$ to objects $F(A)$ in $\mathcal{B}$ and morphisms $f : A_1 \longrightarrow A_2$ in $\mathcal{A}$ to morphisms

$$F(f)\colon F(A_1) \longrightarrow F(A_2).$$

A contravariant functor reverses the arrows, i.e., it holds

$$F(f)\colon F(A_2) \longrightarrow F(A_1).$$

An example is the covariant functor $F\colon \mathbf{Top} \longrightarrow \mathbf{Set}$, which sends a topological space to the underlying set and continuous maps to the underlying map, i.e., it ignores continuity. Other examples are the covariant functors of homology and homotopy groups, which represent functors $\mathbf{Top} \longrightarrow \mathbf{Gr}$ in the category of groups. At the level of the underlying chain complexes of abelian groups, which Emmy Noether already considered, the functor $\mathbf{Top} \longrightarrow \mathbf{Ch}$ can be considered, which assigns to each topological space X the singular chain complex $\mathbb{Z}\mathrm{Sing}_.(X)$.

Functors can be compared with each other. A natural transformation T between two functors F and G assigns to each object A of $\mathcal{A}$ a morphism $T_A\colon F(A) \longrightarrow G(A)$, so that the depicted diagram in the category $\mathcal{B}$ commutes.

$$
\begin{array}{ccc}
F(A) & \xrightarrow{\;\;F(f)\;\;} & F(B) \\
\Big\downarrow{\scriptstyle T_A} & & \Big\downarrow{\scriptstyle T_B} \\
G(A) & \xrightarrow[\;\;G(f)\;\;]{} & G(B)
\end{array}
$$

Natural transformation.

Even with categories there is the basic question of equality and there can be equivalent variants. Equivalences of two categories $\mathcal{C}$ and $\mathcal{D}$ arise through two functors

$$F\colon \mathcal{C} \longrightarrow \mathcal{D} \text{ and } G\colon \mathcal{D} \longrightarrow \mathcal{C},$$

so that there are natural transformations

$$\varepsilon\colon \mathrm{id}_{\mathcal{D}} \xrightarrow{\;\cong\;} F \circ G \text{ and } \eta\colon G \circ F \xrightarrow{\;\cong\;} \mathrm{id}_{\mathcal{C}}$$

which induce isomorphisms on all objects.

Presheaves and Sheaves

With the help of functors, sheaves and presheaves can be defined as new objects on topological spaces. Such objects are important as structure sheaves and as coefficients for cohomology theories. In the history of the development of mathematics, they were introduced around 1950 within complex analysis by Henri Cartan and Kiyoshi Oka in the form of sheaves of holomorphic functions on complex-analytical spaces.[198]

A presheaf $\mathcal{F}$ is a contravariant functor from the category of open sets $\mathbf{Off}(X)$ in a topological space X with values in the category $\mathbf{Set}$. This means that for two open sets $V \subset U$ there is a restriction mapping

$$\rho_V^U : \mathcal{F}(U) \longrightarrow \mathcal{F}(V)$$

which must satisfy two functor properties. Firstly, the trivial inclusion $U \subset U$ induces the identity $\mathrm{id}_{\mathcal{F}(U)}$ as restriction mapping ρ_U^U and secondly, for three nested open sets $W \subset V \subset U$ the composition rule for restriction mappings applies (see diagram). The category of presheaves is denoted by $\mathbf{Psh}(X)$.

$$\mathcal{F}(U) \xrightarrow{\ \rho_V^U\ } \mathcal{F}(V) \xrightarrow{\ \rho_W^V\ } \mathcal{F}(W)$$
$$\rho_W^U = \rho_W^V \circ \rho_V^U$$

Functor property of the restriction mappings.

Sheaves of sets on a topological space X are special presheaves that fulfil an additional property.[199] They form a category, denoted by $\mathbf{Sh}(X)$. A sheaf or a presheaf forms a family of sets over X, because for every point $x \in X$ the stalk $\mathcal{F}_x$ of the sheaf can be defined as a set.[200]

The category of presheaves $\mathbf{Psh}(X)$ is an example of a functor category. It consists of the collection

$$\hat{\mathcal{C}} = \mathbf{Set}^{\mathcal{C}^{\mathrm{op}}}$$

of all functors $\mathcal{C}^{\mathrm{op}} \longrightarrow \mathbf{Set}$ on a given category $\mathcal{C}$. The category $\mathbf{Set}$ is the special case of the presheaves $\mathbf{Psh}(*)$ on the point. For the category $\mathcal{C} = \Delta.$ of finite ordered sets of the form

$$\Delta_n = \{0 < 1 < \cdots < n\}$$

with monotone mappings as morphisms, the functor category $\hat{\Delta}.$ is the category $\mathbf{sSet}$ of simplicial sets.[201]

Functor categories often have better properties than $\mathcal{C}$ itself and every small category possesses a useful embedding

$$y : \mathcal{C} \longrightarrow \hat{\mathcal{C}}, \ B \mapsto \mathrm{Hom}_{\mathcal{C}}(-, B),$$

which is also called Yoneda embedding[202] after Nobuo Yoneda.

The Category **Set** as Elementary Topos

How can **Set** be defined as a category without using axiomatic set theory? William Lawvere has in the 1960s described the category of sets axiomatically in a structural way which provides an alternative to Zermelo–Fraenkel set theory. Lawvere has named his approach the "Elementary theory of the category of sets" (ETCS). Later, other versions such as SEAR ("Sets, elements and relations") by Jeremy Avigad emerged.[203]

In such categories $\mathcal{E}$, the element symbol is not used. Instead, the elements of M are given by arrows

$$1 \xrightarrow{x} M$$

from a canonical terminal object 1 to M. Terminal means that for all objects A in $\mathcal{E}$ a morphism $A \longrightarrow 1$ exists. As one of the defining axioms for an elementary topos, it is required that two morphisms $f, g: M \longrightarrow M'$ are equal in $\mathrm{Hom}(M, M')$ if and only if their composition with all elements $1 \xrightarrow{x} M$ coincides. It is thus required that sufficiently many elements are present to guarantee the function extensionality of morphisms.

All further axioms of ETCS can be expressed by the fact that **Set** is an elementary topos that has an object $\mathbb{N}$ of natural numbers. This concept was coined by William Lawvere and Myles Tierney around 1970, because they recognised that certain properties of **Set** in the form as Lawvere demanded for ETCS form an important class of categories.

The first group of properties of an elementary topos states that $\mathcal{E}$ is a cartesian closed category. Firstly, for any two objects A, B in $\mathcal{E}$ the binary product $A \times B$ exists (see diagram).

$$
\begin{array}{ccc}
A \times B & \xrightarrow{\ \mathrm{pr}_B\ } & B \\
{\scriptstyle \mathrm{pr}_A} \downarrow & & \downarrow \\
A & \xrightarrow{\hspace{2cm}} & 1
\end{array}
$$

Binary product.

The binary product fulfils the universal property that every morphism

$$T \xrightarrow{h} A \times B$$

is uniquely given by two morphisms $T \xrightarrow{f} A$ and $T \xrightarrow{g} B$ into the two components in the form $h = (f, g)$.

Furthermore, there should be an internal exponential object X^B in $\mathcal{E}$, which represents the functions from B to X. This means that all morphisms $B \longrightarrow X$ are represented by morphisms $1 \longrightarrow X^B$ into an object X^B and more generally every morphism

$$A \times B \xrightarrow{\;f\;} X$$

corresponds to a unique morphism

$$A \xrightarrow{\;F\;} X^B$$

where $f(a, b) = F(a)(b)$ holds. This implies that morphisms with multiple arguments can successively be replaced by simple morphisms. In the literature, the exponential object is also called an internal hom-object. The property of being a cartesian closed category holds even locally in an elementary topos. This means that all comma categories $\mathcal{E}/_X$ are cartesian closed. Their objects A are defined as objects in $\mathcal{E}$ along with a morphism $A \longrightarrow X$.

In addition to the terminal object 1, which corresponds to a singleton set in **Set**, there is an initial object 0, which corresponds to the empty set in **Set**. Initial means that there are morphisms $0 \longrightarrow A$ for all A in $\mathcal{E}$. The objects 0 and 1 do not represent numbers or truth values, although a connection exists.

$$
\begin{array}{ccc}
A \times_X B & \xrightarrow{\;\mathrm{pr}_B\;} & B \\
{\scriptstyle \mathrm{pr}_A}\big\downarrow & & \big\downarrow{\scriptstyle g} \\
A & \xrightarrow[\;f\;]{} & X
\end{array}
$$

Pullback diagram.

The second group of properties of an elementary topos requires that the category $\mathcal{E}$ has so-called finite limits and finite colimits, denoted by lim and colim. The simplest example of a limit is the binary product $A \times B$ in **Set**. A more general limit is a pullback object $A \times_X B$ for two morphisms $f : A \longrightarrow X$ and $g : B \longrightarrow X$ in $\mathcal{E}$. The object $A \times_X B$ is the binary product in the comma category $\mathcal{E}/_X$ (see diagram). The object $A \times_X B$ has the universal property that for two morphisms $T \longrightarrow A$ and $T \longrightarrow B$, whose compositions with f and g coincide as morphisms $T \longrightarrow X$, an arrow

$$T \longrightarrow A \times_X B$$

exists, whose respective compositions with the two projections again coincide with the given morphisms. In the case of the terminal object $X = 1$ the binary product results in

$$A \times_X B = A \times B.$$

A special case of a colimit is a pushout object $A +_X B$ for two arrows $f \colon X \longrightarrow A$ and $g \colon X \longrightarrow B$ (see diagram).

$$
\begin{array}{ccc}
X & \xrightarrow{\ f\ } & A \\[4pt]
{\scriptstyle g}\downarrow & & \downarrow{\scriptstyle i_A} \\[4pt]
B & \xrightarrow[\ i_B\ ]{} & A +_X B
\end{array}
$$

Pushout diagram.

In the case of the initial object $X = 0$, the binary sum $A + B$, also called coproduct, results. In **Set**, the coproduct corresponds to the disjoint union.

The third group of properties of an elementary topos demands that the category $\mathcal{E}$ has a subobject classifier Ω together with an arrow $1 \xrightarrow{\ t\ } \Omega$ (t as in true). This guarantees for every subobject $B \subset A$ a classifying arrow χ, so that in the depicted diagram t becomes a universal embedding.

$$
\begin{array}{ccc}
B & \longrightarrow & 1 \\[4pt]
\downarrow & & \downarrow{\scriptstyle t} \\[4pt]
A & \xrightarrow{\ \chi\ } & \Omega
\end{array}
$$

Subobject classifier.

The idea here is that Ω is a collection of truth values and contains a distinguished element 1 that corresponds to the value true. In the example of the elementary topos **Set**, $\Omega = \{0, 1\}$ with the usual classical truth values 0 (false) and 1 (true). The morphism χ in this example is the characteristic function of the subset B, i.e., for all elements a of A it holds

$$
\chi(a) = \begin{cases} 1 & \text{if } a \text{ is an element of } B \\ 0 & \text{otherwise.} \end{cases}
$$

An elementary topos possesses power objects $\mathrm{Pow}(A)$ for every A in $\mathcal{E}$. Thus, the subobject classifier Ω is of the form $\mathrm{Pow}(1)$. All power objects $\mathrm{Pow}(A)$ in $\mathcal{E}$ are internal Heyting algebras and all lattices

$$
\mathrm{Sub}_{\mathcal{E}}(A) = \mathrm{Hom}_{\mathcal{E}}(A, \Omega)
$$

of subobjects of an object A in $\mathcal{E}$ possess the structure of external Heyting algebras, which in both cases are usually not Boolean algebras.[204] The operation $\wedge$ on Ω is defined by the universal property as shown in the diagram below.

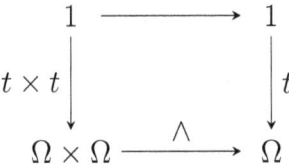

Intersection operation ∧ in the Heyting algebra Ω.

The operations $\wedge$ and $\vee$ correspond to finite products and finite coproducts in the elementary topos. The operation $\Rightarrow$ maps two objects A, B in Ω to the exponential object B^A. With this operation, or with the inequality $\leq$, Ω can be considered as a category, in which there exists a unique or no arrow between two objects A, B. It is usually additionally required that an elementary topos has an object $\mathbb{N}$ of natural numbers, i.e., a null object 0 and a successor morphism $S\colon \mathbb{N} \longrightarrow \mathbb{N}$.[205]

There are, in addition to **Set**, two further examples of elementary toposes that we have already encountered, namely the functor categories

$$\hat{\mathcal{C}} = \mathbf{Set}^{\mathcal{C}^{\mathrm{op}}}$$

for a category $\mathcal{C}$ and the category $\mathbf{Sh}(X)$ of sheaves of sets on a topological space X. In all these examples, Ω consists of a certain set of sheaves.[206] The three mentioned types of elementary toposes are so-called Grothendieck toposes, which Alexander Grothendieck introduced to define new topologies in algebraic geometry.[207] The category **Top** on the other hand is not even an elementary topos, because in it there are no exponential objects.[208]

2-Categories and Bicategories

Higher categories and infinity categories are modifications of categories that can handle the equivalence of mathematical objects better than ordinary categories.[209] In this process, categorical calculation rules are weakened in different ways and the coherence of this approach is secured by higher morphisms (see diagram).

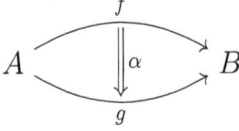

Higher 2-morphism α between two 1-morphisms.

There are precise definitions for higher n-categories for small natural numbers $n = 2, 3, \ldots$, which in the limit case $n \to \infty$ are called infinity categories. We start by explaining the case of a strict 2-category. In them, there are additional 2-morphisms between the ordinary 1-morphisms. The composition of arrows can occur in different ways. Thus, in a 2-category, in addition to the composition of two ordinary morphisms, there is a vertical and a horizontal possibility to link the 2-morphisms (see diagram).

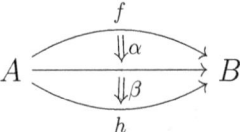

Vertical composition.

In a strict 2-category, it is required that both types of compositions are associative and commute with each other. Thus, the diagram shown has no ambiguities. In addition, there should be 2-morphisms id_f for each morphism f that are identities with respect to both compositions.

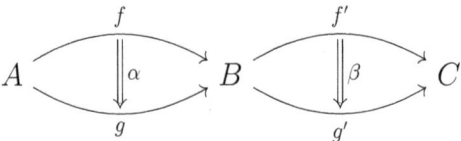

Horizontal composition.

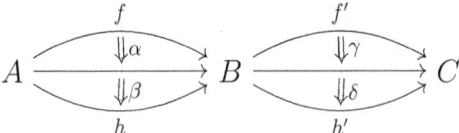

Both compositions commute with each other.

The most important example of a strict 2-category is **Cat**, the category of all (small) categories with functors as morphisms and natural transformations as 2-morphisms. Another example is the category **Gr** of groups, where the morphisms are the group homomorphisms

$$G \xrightarrow{\ f\ } H, \text{ with } f(ab) = f(a)f(b)$$

and the 2-morphisms

$$f \Rightarrow f'$$

between f and f' are given by the conjugations with elements of the group:

$$f'(g) = hf(g)h^{-1}, \text{ with } h \in H.$$

As a rule, strict 2-categories are rare, although it can be shown that every 2-category is equivalent to a strict one. More often, bicategories are considered. These are weakenings of 2-categories, in which the composition of 1-morphisms only needs to be associative modulo certain higher expressions.

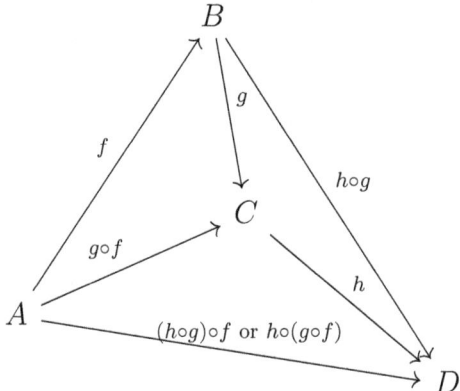

Tetrahedron for the associator.

As a replacement for the associative law, there is an invertible 2-morphism

$$a_{h,g,f} : (h \circ g) \circ f \Rightarrow h \circ (g \circ f)$$

—called the associator—which depends naturally on h, g and f. It is represented by a 3-dimensional tetrahedron, the sides of which correspond to the possible compositions (see diagram). Associators for four objects i, h, g, f satisfy the pentagon rule, which states that the two ways of going around the pentagon yield the same result (see diagram).

Additionally, in a bicategory, the existence of invertible 2-morphisms

$$l_f : \mathrm{id}_B \circ f \Rightarrow f, \quad r_f : f \circ \mathrm{id}_A \Rightarrow f$$

for every morphism $f : A \longrightarrow B$ is required, which are called unitors.

Higher Categories and Infinity Categories

The situation becomes even more interesting with 3-categories, where the pentagon rule is not a equality, but is replaced by a 3-morphism, which is called the associahedron. In general, there are n-categories for all n. In these, higher k-morphisms are often described in a

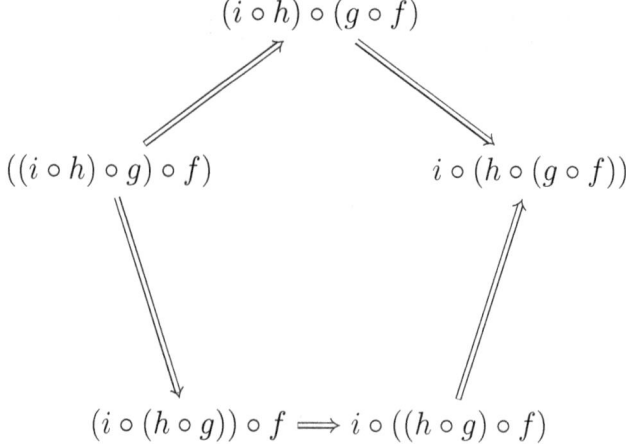

Pentagon rule.

geometric way that reminds one of simplicial structures. The k-dimensional morphisms correspond to k-dimensional simplices, which describe the relationships between $(k-1)$-dimensional objects. For example, the composition $g \circ f$ of two morphisms f and g is described by the choice of a 2-dimensional simplex Δ_2 (see diagram).

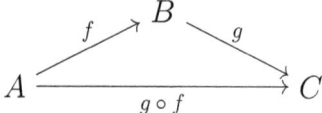

Composition of morphisms.

Infinity categories are even more general than n-categories. So far, no clear axiomatic system has prevailed for them. In the literature, primarily simplicial models of higher categories and infinity categories are studied. It seems desirable to define them intrinsically without resorting to models.[210]

Infinity categories have objects and higher k-morphisms for all $k \geq 1$. The calculation rules and relationships between higher morphisms are, as in the cases of 2- and 3-categories or bicategories, to be determined. Thus, the composition of higher morphisms need not be unique nor strictly associative. But the basic principle applies that the choices for non-unique higher morphisms are in a suitable sense equivalent to each other, if they are not unique. Such coherence conditions among k-morphisms are given by $(k+1)$-morphisms. In the simplest case, the equivalence of two morphisms $f, g \colon A \longrightarrow B$ is described by a 2-simplex (see diagram).

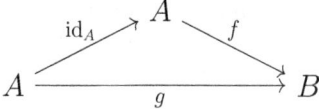

Equivalence of two morphisms f, g.

Infinity categories are in a certain sense the limit case of a hierarchy of (n, r)-categories. The letter n stands for the same counting as in n-categories and the letter r with $0 \leq r \leq n + 1$ for another index. In an (n, r)-category, k-morphisms become trivial if $k > n$, and they become invertible equivalences if $k > r$. Also in infinity categories this classification can be made and (∞, r)-categories result. Most often in this context, $(\infty, 1)$-categories are meant when we speak of infinity categories. Overall, the following scheme emerges, in which the concept of the set (or class) and the partially ordered set (poset) generalise:

	$n = 0$	$n = 1$	$n = 2$	$\cdots$	$n = \infty$
$r = 0$	set	groupoid	2-groupoid	$\cdots$	$(\infty, 0)$-category
$r = 1$	poset	category	$(2, 1)$-category	$\cdots$	$(\infty, 1)$category
$r = 2$	poset	2-poset	2-category	$\cdots$	$(\infty, 2)$-category
$r = 3$	poset	2-poset	3-poset	$\cdots$	$(\infty, 3)$-category
$\vdots$	$\vdots$	$\vdots$	$\vdots$	$\ddots$	$\vdots$

The most important example of a normal category which also exhibits the characteristics of an infinity structure is the category **Top**. In it, in addition to the usual objects and 1-morphisms, there are 2-morphisms, the homotopies h between continuous mappings

$$f, g: X \longrightarrow Y,$$

which are defined by a continuous mapping

$$h: [0, 1] \times X \longrightarrow Y.$$

Furthermore, there are homotopies between homotopies etc., resulting in typical characteristics of an infinity category, more precisely an $(\infty, 1)$-category. Formally, the correct setup in which these observations make sense is given by model categories of Daniel Quillen, which we will study later.

Another example of an infinity category is the fundamental infinity groupoid $\Pi_\infty(X)$ of a topological space X, which is a generalisation of the fundamental groupoid $\Pi_1(X)$. This object is a $(\infty, 0)$-category, whose objects consist of the points of X and the morphisms consist of the paths. The 2-dimensional higher objects (or 2-morphisms) are data that encode the homotopies between the paths, which are used for the calculation rules. These are the rules for the inverse path p^{-1}, for the composition of paths and for the associative law, all of which only apply up to homotopy. The 3- and higher-dimensional objects are generated by the homotopies between the homotopies and so on.[211]

Attempts have been made to define (n, r)-toposes or elementary (∞, r)-toposes. Such objects must fulfil similar axioms to ordinary elementary toposes.[212]

Models of Infinity Categories

Models of infinity categories are studied in various variants, the quasi-categories or weak Kan complexes, complete Segal spaces, Segal categories or simplicial categories.[213] As a model for the fundamental infinity groupoid $\Pi_\infty(X)$ of a topological space X with enough real paths, the simplicial set $\mathrm{Sing}.(X)$ is used. For this, we use the fact that $\mathrm{Sing}.(X)$ is created by continuous mappings $\Delta_n \longrightarrow X$ for all n, but that homotopies between paths are actually formed by continuous mappings

$$h \colon [0, 1] \times [0, 1] \longrightarrow X,$$

i.e., by continuous mappings of a square $\square^2$ to X. However, a square can be divided into two triangles by its diagonal. This trick ensures that homotopies between paths and their higher generalisations can be traced back to continuous mappings $\Delta_n \longrightarrow X$.

Kan complexes are named after Daniel Kan and are defined by the existence of fillings in the form of an n-simplex Δ_n for those subsets Λ_n^i of $\partial \Delta_n$ where the i-th side was removed from $\partial \Delta_n$ for $0 \le i \le n$. This condition ensures that all morphisms of dimension $n \ge 1$ are invertible. Therefore, Kan complexes are models of $(\infty, 0)$-categories or infinity groupoids.

The inner subset Λ_2^1 is represented by two morphisms f and g that can be linked (see diagram). The filling is realised by choosing a composition $g \circ f$.

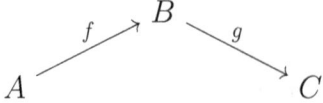

Inner subset Λ_2^1 in $\partial \Delta_2$.

In weak Kan complexes, it is required that only the inner subsets Λ_n^i for $1 \le i \le n - 1$ have fillings. As a result, in weak Kan complexes all morphisms of dimension $n \ge 2$ are invertible and weak Kan complexes are models of $(\infty, 1)$-categories.

Category Theory and the Concept of Space

The development of a most general concept of space and number is a goal within and outside of mathematics. One of the prevailing trends is the categorification of mathematical structures. In particular, $(\infty, 0)$- and $(\infty, 1)$-categories are the crucial structures with which we can generalise the concept of space and number.

As we have seen, the fundamental group $\pi_1(X, *)$ is usefully extended to the infinity groupoid $\Pi_\infty(X)$—an $(\infty, 0)$-category. How did this idea come about? Many recent developments in category theory in mathematics can be traced back to a nearly 600-page manuscript "A la poursuite des champs"[214] by Alexander Grothendieck, which began with a letter to Daniel Quillen dated February 19th 1983. This manuscript greatly influenced the further development of homotopy theory and (higher) category theory. Around the same time, the model categories of Daniel Quillen and the related path categories of Kenneth Brown, both of which are main tools of homotopy theory to work in a structural way in the category **Top**, were created.

Alexander Grothendieck conjectured that every infinity groupoid is equivalent to one of the form $\Pi_\infty(X)$. Assuming this homotopy hypothesis by Grothendieck, it follows that the homotopy types of topological spaces are given by the totality of all infinity groupoids, thus by an $(\infty, 1)$-category, which can be constructed from model categories or path categories.[215] This observation shows that infinity groupoids or $(\infty, 0)$-categories and homotopy theory are closely related.

We have already seen that the usual definition of $\Pi_\infty(X)$ only makes sense for certain topological spaces X. For various interesting situations, a different approach is better to arrive at an infinity groupoid $\Pi_\infty(X)$. In algebraic geometry, the underlying topological spaces almost never have real paths. However, there is an alternative definition of a corresponding simplicial object $\Pi_\infty(X)$, if X is an algebraic scheme, [216] an algebraic space[217] or an algebraic stack.[218] It can be constructed using the étale homotopy theory and suitable étale—especially unbranched—coverings of X.[219]

Gauge theories of mathematical physics often use differentiable stacks, on which gauge transformations operate in the form of groupoids and preserve physical properties, such as the Lagrange functions in Yang–Mills theories.[220] Higher n-stacks, derived algebraic spaces, locales and condensed spaces are even more general space theories, which also play a role in mathematical physics and all of which have connections with infinity structures. The concept of equivalence plays a crucial role in them.[221] Condensed spaces are particularly promising among these variants, as they unite many space concepts. They are defined as functors

$$T : \textbf{Profinite Sets}^{\text{op}} \longrightarrow \textbf{Set},$$

which satisfy the sheaf condition with respect to a suitable topology. Every topological space can be represented as a condensed space. Profinite sets are limits of so-called directed systems of finite sets. Examples include the p-adic integers $\mathbb{Z}_p = \varprojlim \mathbb{Z}/p^n\mathbb{Z}$, which are the limit of the finite sets $\mathbb{Z}/p^n\mathbb{Z}$ for $n \geq 1$, or the numbers $\hat{\mathbb{Z}} = \varprojlim \mathbb{Z}/m\mathbb{Z}$ as the limit of all sets $\mathbb{Z}/m\mathbb{Z}$ for $m \geq 1$.

If we want to study topological spaces X not only up to homotopy but up to homeomorphism, it is advisable to replace the space X with the elementary topos **Sh**(X), because X can be reconstructed with it, provided X is a Hausdorff space.[222] Even more flexible is the infinity topos **Sh**$_\infty(X)$ in the form of an $(\infty, 1)$-category of infinity sheaves of infinity groupoids on X. Such generalised spaces can, despite their degree of abstraction, be used as efficient replacements for classical topological spaces and yet preserve the properties of traditional space concepts.

All different types of generalised spaces can be assigned homology groups and other invariants. We have already encountered the singular homology groups

$H_n(X, \mathbb{Z})$ with integer coefficients of a topological space or a manifold X. Such homology groups, also those with coefficients other than $\mathbb{Z}$, contain primary invariants of X and of geometric objects on X, such as the Chern classes of X or of sheaves $\mathcal{E}$ on X. These invariants express geometric curvature properties of X or of $\mathcal{E}$ and thereby classify these objects in a certain way. Refinements of the homology groups, such as the differential homology or the Deligne homology, allow the definition of secondary invariants, which refine the primary invariants and describe some properties of spaces and sheaves more precisely. In physics, Chern classes and their generalisations play an important role. The simplest example of this is the magnetic flux in Maxwell's theory, which can be expressed as Chern class c_1 of a $U(1)$-bundle, as Paul Dirac discovered.

There are algebraic versions of the homology groups and the homotopy groups of algebraic schemes, which can be best understood through the $\mathbb{A}^1$-homotopy theory discovered by Vladimir Voevodsky and others. A well-known example among these are the classical Chow groups $CH_i(X)$ of a scheme X. Such groups are also called motivic homology groups and homotopy groups, because the motives imagined by Grothendieck are abstractions of the homological properties of algebraic schemes. Voevodsky proved conjectures of arithmetic, such as the Milnor conjecture and the Bloch–Kato conjecture, using the $\mathbb{A}^1$-homotopy theory, and was awarded the Fields Medal for this in 2002. Such results connect periods, motives and other invariants of algebraic objects with number-theoretical questions, partly in generalisation of the Riemann hypothesis.[223]

Type Theory

8

From the discovery of the antinomies of set theory, Bertrand Russell concluded that a hierarchy of mathematical objects is necessary. As a solution, he invented type theory. His ideas were incorporated into the book "Principia mathematica" with Alfred North Whitehead. The presentation therein seemed complicated and initially did not catch on. Leon Chwistek and Frank P. Ramsey attempted to rectify some of these problems and created the simple type theory. However, it was an article by Alonzo Church that made this approach widely known using the λ-calculus.[224]

Per Martin–Löf has further developed the simple type theory from around 1971 and built the intuitionistic dependent type theory as a possible foundation of mathematics alongside set theory and category theory. He has also written philosophical works in which he linked Kant's theory of synthetic judgements with type theory and Gödel's incompleteness theorems.[225] A first-order fragment of dependent type theory was developed by Mihály Makkai around 1995.

The concept of equality plays a significant role in dependent type theory. In the initial versions, this was extensionally[226] formulated and thus narrowly defined. Later versions included an intensional identity type, which was Martin–Löf's most important and influential idea. The pair of concepts intensional versus extensional is related to Frege's distinction between sense and meaning.

Since the beginning of our century, the theory of Martin–Löf has been extended to homotopy type theory (alias univalent foundations), which gives dependent type theory a homotopically motivated topological interpretation. This fits well with the character of Martin–Löf's type theory, leads to a more intuitive view and is an inspiration for the further development of type theory. Another consequence of Martin–Löf's type theory and its variants is the possibility of machine verification of proofs. This opens up interesting future perspectives for mathematics and its applications.[227]

S. Müller-Stach, *The Code of Mathematics*, Mathematics Study Resources 11, https://doi.org/10.1007/978-3-662-69483-1_8

Martin–Löf Type Theory

A fundamental notation of type theory forms judgements of the form

$$a : A.$$

By this we understand a term a that belongs to the type A. Two terms $a, b : A$ are called definitionally equal, denoted $a \equiv b$, if the designation b is just an alternative designation of a. This form of definitional equality is the closest possible form of equality in type theory.

In contrast to set theory, the judgement $a : A$ avoids intensional constructions in type theory. If a term a is given, then is usually also the associated type A that a inhabits, i.e., each term a has a well-defined type. In set theory this is quite different, because an element a can be an element of several sets. The set-theoretical operations $\cap$ and $\cup$ are obviously intensional. Therefore, they are not present in this form in type theory.

There are important elementary types, such as the empty type **0**, which is not inhabited, or the single-element type **1**, which is inhabited by exactly one term. In addition, there is the two-element Boolean type **Bool**, whose terms are the classical truth values $\top$ (true) and $\bot$ (false).

The most important basic type construction for any two types A, B is the extensional function type

$$(A \longrightarrow B)$$

which consists of functions or mappings $f : A \longrightarrow B$. This is a primitive type, i.e., it is not definable from the remaining structures of the theory. In set theory, the function type is not primitive, because functions can be defined by their graph as a subset. Mappings $f : A \longrightarrow$ **Bool** are also called predicates.

Each mapping f corresponds to a rule, with which for a given $a : A$ an $f(a) : B$ can be obtained. If two mappings $f, g : A \longrightarrow B$ in type theory agree on every function value, they are not necessarily equal. Such a requirement, called function extensionality and valid in set theory, does not follow a priori from type theory, but only with additional assumptions such as the univalence axiom by Voevodsky. In the intuitionistic version of type theory, the mapping rule is constructively given. In fact, f in the case $A = B = \mathbb{N}$ under certain conditions is a computable function.[228]

With the help of the function type, types can be related to each other. For example, there is a sequence of mappings between number systems[229]

$$\mathbb{N} \longrightarrow \mathbb{Z} \longrightarrow \mathbb{Q} \longrightarrow \mathbb{R} \longrightarrow \mathbb{C} \longrightarrow \mathbb{H} \longrightarrow \mathbb{O},$$

under which the term $5 : \mathbb{N}$ is sent to its respective images $5_{\mathbb{Z}}, 5_{\mathbb{Q}}, 5_{\mathbb{R}}$ etc., which we also denote by 5 for simplicity.

Another application of this type is the possibility of defining a power type as function type

$$\mathrm{Pow}(A) = (A \longrightarrow \textbf{Bool}).$$

This classifies all decidable subtypes B of A that correspond to a predicate $\chi : A \longrightarrow$ **Bool**, which takes the value true on B.

As in set theory, there is the type of the (binary) product

$$A \times B$$

and the (binary) sum

$$A + B.$$

In the former case, the terms are given by pairs (a, b) with $a : A$ and $b : B$, in the latter case a term $c : A + B$ either fulfils $c : A$ or $c : B$.

Universes and Families of Types

As in set theory, the impredicative assumption of a type of all types creates a paradox analogous to Russell's antinomy, as noted by Jean–Yves Girard. Martin–Löf and others have developed predicative and impredicative paradox-free versions over the past 50 years using type-theoretical universes. Grothendieck universes were introduced by Alexander Grothendieck to address set-theoretical problems in category theory. They are based on the assumption of strongly inaccessible cardinal numbers.[230]

The universes U postulated in type theory themselves form types that are inhabited by certain types $A : U$ as terms. Universes are often iterated in a given type theory in ascending order of size in a finite or even infinite sequence

$$U : U' : U'' : U''' : \cdots$$

each representing a term in the next universe, i.e., U is a type within the universe U', this in turn in U'' and so on. Useful properties of universes U are that they can first define families of types and secondly types become terms $A, B : U$ and can thus be compared within U. Such a sequence of universes generates a high degree of flexibility in type theory. As a rule, at least one additional universe U' is required with $U : U'$ to enable certain type constructions in U.

A family of types $B(x)$ with the parameter $x : A$ is given under these conditions by a function

$$B : A \longrightarrow U,$$

where U is a given universe. Geometrically, a family, as indicated in the figure, is illustrated as a deformation where the types $B(x) : U$ vary depending on $x : A$.

Family of types $B(x)$ for three values of $x : A$.

What can be said about the existence of type-theoretical universes and thus of types? Type theory is a syntactic construct, so the concept of existence for objects in type theories is—at least from a from a nominalistic position—meaningless. The crucial question is consistency, i.e., the contradiction-free nature of such deductive systems. Existence is often discussed in a semantic context in set-theoretical or categorical models. However, in our opinion, even there, existence is a subordinate—if not irrelevant—question compared to consistency. Axioms based on very large transfinite cardinal numbers offer a solution to the questions of consistency and existence. However, this answer shifts the problem to the question of consistency in superordinate deductive systems that can describe such numbers, and does not necessarily make it easier.

Dependent Types

The function type can be significantly generalised, leading to the concept of dependent types. The dependent function type

$$\prod_{x:A} B(x)$$

is given by a family of types $B(x)$, i.e., by a function

$$B : A \longrightarrow U$$

into a given universe U with $B(x) : U$. A term

$$s : \prod_{x:A} B(x)$$

in the dependent function type is a generalised mapping s that assigns a $s(x) : B(x)$ to each $x : A$, but otherwise has similar calculation rules as mappings. It is occasionally also denoted as

$$s = \lambda x.s(x)$$

using Church's λ-calculus. If $B(x) = B$ is a constant family, the special case of the mapping type

$$\prod_{x:A} B = (A \longrightarrow B)$$

arises. The dependent pair type

$$\sum_{x:A} B(x)$$

arises from a family $B(x)$ as a generalisation of the product type $A \times B$. Terms in it are pairs (x, y) with $y : B(x)$ for an $x : A$. There is a canonical projection mapping

$$\text{pr}: \sum_{x:A} B(x) \longrightarrow A$$

to the first factor. The binary product type

$$\sum_{x:A} B = A \times B$$

arises in the special case that $B(x) = B$ is constant.

Identity Types

An important new type is the identity type, denoted by

$$\text{Id}_A(a, b) \quad \text{or with} \quad (a =_A b).$$

It is a primitive type like the mapping type and depends on the respective type theory. If it is inhabited by a term p, p expresses a propositional equality between a and b. This is a weakening of the definitional equality $a \equiv b$, which expresses that a and b denote the same term. We imagine the terms $p : \text{Id}_A(a, b)$ as paths p from a to b. The definitional equality corresponds to the constant path.

The identity type is an example of an inductive type, of which we will get to know more. It represents the concept of polymorphism from theoretical computer science, because this type does indeed depend on the type A, but does so in a universal way. The type $\text{Id}_A(a, a)$ is always inhabited by a canonical term refl_a, the constant path. The identity type Id_A without a given pair (a, b) can be considered as a dependent type

$$\text{Id}_A = \sum_{(a,b):A \times A} \text{Id}_A(a, b) \longrightarrow A \times A$$

and is called the path space of A. As in topology, the constant path refl_a, the concatenation of paths $q \circ p$ and the inverse path p^{-1} in Martin–Löf–type theory fulfil the following relationships by means of suitable mappings

$$\mathbf{1} \longrightarrow \text{Id}_A(a, a)$$
$$\text{Id}_A(a, b) \times \text{Id}_A(b, c) \longrightarrow \text{Id}_A(a, c)$$
$$\text{Id}_A(a, b) \longrightarrow \text{Id}_A(b, a).$$

The formulas

$$p^{-1} \circ p = \text{refl}_a$$
$$p \circ p^{-1} = \text{refl}_b$$
$$(p \circ q) \circ r = p \circ (q \circ r)$$

do not hold under definitional equality, but only when using the twice iterated identity type. This kind of higher conditions forms the structure of an infinity groupoid, which is associated with each type A.[231]

Inductive Types

We will now introduce the examples $\mathbb{N}$, $\mathbb{S}^1$ and $\mathbb{I}$ of inductive types, which have already appeared in different forms in previous chapters. They are, so to speak, born out of pure logic and are determined by their calculation rules and their universal properties. There are inductive types in an even more general form, sometimes referred to as W-types or higher inductive types.

Let's start with the type of natural numbers $\mathbb{N}$. This is a recursive type, given by the constructor

$$\mathbb{N} = \begin{cases} 0 : \mathbb{N} \\ S : \mathbb{N} \longrightarrow \mathbb{N} \quad \text{(successor function)}. \end{cases}$$

Similar to Dedekind, this type only provides the initial element 0 and the successor function S. The natural numbers are then given by the infinite sequence

$$0, 1 = S(0), 2 = S(S(0)), \ldots$$

Another example is the circle $\mathbb{S}^1$, which is introduced by

$$\mathbb{S}^1 = \begin{cases} 0 : \mathbb{S}^1 \quad \text{(base point)} \\ \text{loop} : \text{Id}_{\mathbb{S}^1}(0, 0) \quad \text{(loop generator)}, \end{cases}$$

i.e., by postulating a base point and a loop that define the circle. It should be noted that this circle only contains one point 0. Similarly, the interval $\mathbb{I}$ can be introduced by

$$\mathbb{I} = \begin{cases} 0, 1 : \mathbb{I} & \text{(start and end point)} \\ \text{path} : \text{Id}_\mathbb{I}(0, 1). \end{cases}$$

The order here is reversed compared to topology, where the real interval $[0, 1]$ defines the topological path from 0 to 1. This is because paths are primitive objects in type theory. Both $\mathbb{S}^1$ and $\mathbb{I}$ correspond—due to their few inhabitants—as types not to a real circle or interval, but are synthetic constructions thereof. An existential statement for these three types $\mathbb{N}$, $\mathbb{S}^1$ and $\mathbb{I}$ is not intended by their definition, but they are free inductive constructions, the consistency of which is the crucial question. The constructors of inductive types suggest uniqueness due to their simplicity. For example, the constructor for $\mathbb{N}$ essentially agrees with Dedekind's definition of natural numbers. The simple logical construction idea and the uniqueness are in line with logicism and the Platonic world of ideas.[232]

Type Theory as a Deductive System

Type theories can be understood as deductive systems with suitable judgements and inference rules. These are closely related to the formalism of the so-called calculus of inductive constructions by Thierry Coquand and Gérard Huet.[233] The basis for this are contexts Γ, which consist of a series of terms

$$x_0 : A_0, x_1 : A_1(x_0), \ldots, x_n : A_n(x_0, \ldots, x_{n-1}),$$

where the A_i are dependent types and the x_i are variables that stand for terms to be substituted. The contexts of a type theory T form the syntactic category $\mathcal{S}_T$, which is naturally associated with a dependent type theory. Their morphisms arise from obvious mappings between contexts, which go back to arrows between types.[234]

The most fundamental judgements in a type theory with a given universe U are depicted in the table when using contexts Γ.

$\Gamma \vdash A : U$	type
$\Gamma \vdash a : A$	term
$\Gamma \vdash a \equiv_A b$	definitional equality of terms
$\Gamma \vdash A \equiv_U B$	definitional equality of types

Basic judgements of type theory.

In addition to these elementary cases, judgements can arise from four sorts of rules, which we want to present in a simplified form in the following. There are formation rules for the generation of types, introduction and elimination rules for terms in types as well as computation rules for the substitution of arguments in type constructions. Some of these rules are known in type theory under the names α-conversion (renaming of bound variables), η-conversion (function extensionality) and β-reduction (substitution of arguments).

These four rules are described in the case of the dependent type $\prod_{x:A} B(x)$ in the following list using the λ-calculus:

$$\Pi - \text{form:} \qquad \frac{\Gamma \vdash A : U \quad \Gamma, x : A \vdash B(x) : U}{\Gamma \vdash \prod_{x:A} B(x) : U}$$

$$\Pi - \text{intro:} \qquad \frac{\Gamma, x : A \vdash b(x) : B(x)}{\Gamma \vdash b = \lambda x.b(x) : \prod_{x:A} B(x)}$$

$$\Pi - \text{elim:} \qquad \frac{\Gamma \vdash s : \prod_{x:A} B(x) \quad \Gamma \vdash a : A}{\Gamma \vdash s(a) : B[\frac{a}{x}]}$$

$$\Pi - \text{comp:} \qquad \frac{\Gamma, x : A \vdash b(x) : B(x) \quad \Gamma \vdash a : A}{\Gamma \vdash (\lambda x.b(x))(a) \equiv b[\frac{a}{x}] : B[\frac{a}{x}]}.$$

In the case of the inductive identity type, the following rules apply, where $C(a, b, p)$ is a type dependent on $a, b : A$ and $p : \text{Id}_A(a, b)$ and contains terms of the form $c(a, b, p) : C(a, b, p)$:

$$\text{Id} - \text{form:} \qquad \frac{\Gamma \vdash A : U \quad \Gamma \vdash a : A \quad \Gamma \vdash b : A}{\Gamma \vdash \text{Id}_A(a, b) : U}$$

$$\text{Id} - \text{intro:} \qquad \frac{\Gamma \vdash a : A}{\Gamma \vdash \text{refl}_a : \text{Id}_A(a, a)}$$

$$\text{Id} - \text{elim:} \qquad \frac{\Gamma \vdash a, b : A \quad \Gamma \vdash p : \text{Id}_A(a, b) \quad \Gamma \vdash c_a : C(a, a, \text{refl}_a)}{\Gamma \vdash \text{ind}_A(a, b, p, c_a) : C(a, b, p)}$$

$$\text{Id} - \text{comp:} \qquad \frac{\Gamma \vdash a : A \quad \Gamma, z : A \vdash c(z) : C(z, z, \text{refl}_z)}{\Gamma \vdash \text{ind}_A(a, a, \text{refl}_a, c[\frac{a}{z}]) \equiv c[\frac{a}{z}] : C(a, a, \text{refl}_a)}.$$

The third rule is the path induction rule by Martin–Löf, which represents a far-reaching generalisation of Leibniz's invariance rule.

The inductive type $\mathbb{N}$ of natural numbers has the following rules, where $C(x)$ is a type dependent on $x : \mathbb{N}$:

$$\mathbb{N} - \text{form:} \qquad \frac{}{\vdash \mathbb{N}}$$

$$\mathbb{N} - \text{intro:} \qquad \frac{}{\vdash 0 : \mathbb{N}} \quad \frac{\Gamma \vdash n : \mathbb{N}}{\Gamma \vdash S(n) : \mathbb{N}}$$

$$\mathbb{N} - \text{elim:} \qquad \frac{\Gamma \vdash c_0 : C[\frac{0}{x}] \quad \Gamma, x : \mathbb{N}, y : C \vdash c_{S(x)} : C[\frac{S(x)}{x}] \quad \Gamma \vdash n : \mathbb{N}}{\Gamma \vdash \text{ind}_{\mathbb{N}}(c, n) : C[\frac{n}{x}]}$$

$$\mathbb{N} - \text{comp}_0 : \qquad \frac{\Gamma \vdash c_0 : C[\frac{0}{x}] \quad \Gamma, x : \mathbb{N}, y : C \vdash c_{S(x)} : C[\frac{S(x)}{x}]}{\Gamma \vdash \text{ind}_{\mathbb{N}}(c, 0) \equiv c_0 : C[\frac{0}{x}]}$$

$$\mathbb{N} - \text{comp}_S : \qquad \frac{\Gamma \vdash c_0 : C[\frac{0}{x}] \quad \Gamma, x : \mathbb{N}, y : C \vdash c_{S(x)} : C[\frac{S(x)}{x}] \quad \Gamma \vdash n : \mathbb{N}}{\Gamma \vdash \text{ind}_{\mathbb{N}}(c, S(n)) \equiv c_{S(n)} : C[\frac{S(n)}{x}]}.$$

The conclusion rules for the remaining types can be found in the literature.[235]

Type Theory and Higher-Order Logic

Type theory is referred to as higher-order logic, as the internal logic of a given type theory generalises first-, second-, third-order logic and so on. In second-order logic, the quantifiers, unlike in first-order logic, can range over predicate variables φ and thus over subsets

$$A = \{x \mid \varphi(x)\}$$

of domains of individuals, which are traversed by the variable x.[236] Third- and higher-order logics generalise this to more general iterations of such situations.

Type theory includes higher logic through the Curry–Howard or also Curry–Howard–Lambek correspondence. It is described with the expression "propositions are types", because it equates types and propositions. Each logical proposition P is understood as the type P of its proofs. For example, the proposition $a =_A b$ is provable if a $p : \mathrm{Id}_A(a, b)$ can be determined by the rules. Conversely, each type A is assigned the proposition which expresses the information whether A is inhabited or not. A proposition is a provable theorem if the associated type P is inhabited by a proof $p : P$. This is obviously a constructive concept of provability and $p : P$ is understood as a program according to the maxim "proofs are programs". In this sense, type theory is comparable to the code of a programming language. Each type A can be assigned a reduced type $||A||$, which is called a mere proposition. If A is inhabited, $||A||$ is inhabited by exactly one term, otherwise it is the empty type. Logical conjunctions in statements correspond to type constructions in the illustrated table.[237]

$\bot$	$\top$	$P \vee Q$	$P \wedge Q$	$P \Rightarrow Q$	$\neg P$	$\exists_{a:A} P(a)$	$\forall_{a:A} P(a)$
0	1	$P + Q$	$P \times Q$	$P \to Q$	$P \to 0$	$\sum_{a:A} P(a)$	$\prod_{a:A} P(a)$

The Curry–Howard correspondence.

This correspondence is to be understood in such a way that, for example, the logical conjunction $P \wedge Q$ of two propositions is provable if $P \times Q$ and thus both P and Q are inhabited. The types $P + Q$ and $\sum_{a:A} P(a)$ are not mere propositions, since for example any term $r : P + Q$ retains the irrelevant information whether r lives in P or in Q. Therefore, a more precise correspondence is

$$P \vee Q = ||P + Q|| \text{ and } \exists_{a:A} P(a) = ||\Sigma_{a:A} P(a)||.$$

Another illustrative example is the implication

$$\neg A \wedge \neg B \Rightarrow \neg(A \vee B),$$

which corresponds to the type[238]

$$P : (A \longrightarrow \mathbf{0}) \times (B \longrightarrow \mathbf{0}) \longrightarrow (A + B \longrightarrow \mathbf{0}).$$

In type theory no form of logic needs to be assumed, because the logical concepts are already integrated, as the table suggests.

In recent years, proof assistance systems such as Agda, Coq, Isabelle, Lean and others have been developed, with which proofs and the verification of software with machine support in the intuitionistic Martin–Löf type theory can be demonstrably carried out correctly. Such systems support the work with meaningful commands. The four-colour theorem, the Feit–Thompson theorem, and the Kepler conjecture were formalised with Coq in 2005, 2012, and 2017 respectively. The verification of a mathematical problem by Peter Scholze, called the liquid tensor experiment, was carried out in 2021 by Johan Commelin and others using Lean. Such possibilities motivated Vladimir Voevodsky to engage with type theory because the complexity of mathematical proofs often causes hard-to-detect errors.[239]

Gödel's System T

A small, quantifier-free type theory with interesting applications is Gödel's system T. With its help, he demonstrated the relative consistency of arithmetic in a 1958 article in the journal Dialectica.[240] Gödel believed that T realised the extension of the finite method hoped for by Hilbert and himself and is more intuitively accessible than the well-ordering of ordinal numbers $\alpha < \varepsilon_0$, which implies the consistency of arithmetic.

The system T contains the inductive type of natural numbers $\mathbb{N}$ and is closed under mappings and binary products $A \times B$ including their two projections. Furthermore, T contains a very general recursion rule, so that T represents all computable functions $f: \mathbb{N}^n \longrightarrow \mathbb{N}$ that arise through transfinite recursion over ordinal numbers $\alpha < \varepsilon_0$. The Ackermann–Péter function is an example of a function that is not primitively recursive and yet can be described by a transfinite recursion. As conclusion rules, the rules of the calculus of inductive constructions are required, as long as they apply to T.

The system T allows the so-called Dialectica interpretation of Heyting arithmetic. To this end, Gödel assigned to each arithmetic formula $\varphi(z)$ with free variables z by an inductive process a formula

$$\varphi^D(z) = \exists x \forall y \, \varphi_D(x, y, z),$$

where φ_D is a quantifier-free formula in T. In addition, it holds that for every intuitionistic proof of $\varphi(z)$ in Heyting arithmetic there exists a term $t = t(z)$ in T such that there is a proof of $\varphi_D(t, y, z)$ within T. With the help of double negation, this results in an interpretation of Dedekind–Peano arithmetic in T.

Topological Interpretation

Martin Hofmann and Thomas Streicher discovered around 1994 that Martin–Löf's type theory allows a homotopically motivated topological interpretation with non-trivial identity types, by establishing a connection with groupoids. A few years later, Steve Awodey, Michael Warren, Vladimir Voevodsky and others contributed further building blocks to the development of type theory, resulting in the emergence of homotopy type theory. Types A are interpreted as homotopy types of topological spaces and mappings $A \to B$ as continuous mappings.[241]

The topological interpretation of the dependent pair type $\sum_{x:A} B(x)$ is the homotopy type of a topological space together with a continuous projection mapping

$$\text{pr}: \sum_{x:A} B(x) \longrightarrow A,$$

so that the fibres $\text{pr}^{-1}(x)$ correspond to the types $B(x)$. The dependent function type $\prod_{x:A} B(x)$ can be interpreted as the space of continuous sections of this projection mapping.

The identity type $\text{Id}_A(a, b)$ is interpreted in homotopy type theory as the space of paths between a and b and the type

$$\text{Id}_A = \sum_{(a,b):A \times A} \text{Id}_A(a, b)$$

as the path space of A. Using constant paths and distinguishing between starting and ending points, we get a sequence of arrows

$$A \longrightarrow \text{Id}_A \longrightarrow A \times A.$$

If a starting point a is fixed, the path spaces $\text{Id}_A(a, -)$ correspond to contractible topological spaces, with the homotopy to the point path refl_a running over the parameterisation of each path. This fact explains why the path induction rule is valid. The iterated identity type

$$\text{Id}_{\text{Id}_A(a,b)}(p, q)$$

between two paths $p, q : \text{Id}_A(a, b)$ is interpreted as homotopy between p and q, i.e., as a path in the path space. Further iterations result in the structure of an infinity groupoid, which is associated with the type A. This shows that homotopy type theory is a synthetic theory of infinity groupoids and not of topological spaces. There are so-called cohesive forms of homotopy type theory which attempt to close this gap.

The topological interpretation of Martin–Löf's type theory has applications in homotopy theory itself. This shows in a surprising way the unity of mathematics through the connection between homotopy theory and the foundations. It is important to understand that this interpretation is synthetic and only becomes apparent in topologically oriented semantic models. Dependent type theory can also be defined and used without such notions. However, it seems to be the case that the nature of type theory comes into its own when types are provided with a topological interpretation.

Propositions, Sets and the h-Stratification

Vladimir Voevodsky has defined a hierarchy for types that is ordered by homotopy levels. With the help of iterated identity types, the mere propositions and sets are generalised. He begins by calling a type A at the lowest level $h = -2$ point-like if

$$\sum_{x:A} \prod_{y:A} \mathrm{Id}_A(x, y)$$

is inhabited, i.e., if there is a term $x : A$ such that for all $y : A$ already $x = y$ applies. In the topological interpretation, this property depends continuously on y and consequently $A \simeq \mathbf{1}$, i.e., the mere proposition of the contractibility of A to a point, denoted by $\mathrm{iscontr}(A)$, is fulfilled.[242]

Point-like set centred in x.

In general, we define recursively

A has level $n + 1$:   For all $x, y : A$ has $\mathrm{Id}_A(x, y)$ level n.

Thus, at level $h = -1$, the mere propositions A are defined by the condition

For all $x, y : A$ is $\mathrm{Id}_A(x, y)$ point-like.

It follows that all $a : A$ are propositionally equal to each other and thus either $A \simeq \mathbf{0}$ is uninhabited or $A \simeq \mathbf{1}$ is inhabited and all path spaces $\mathrm{Id}_A(x, y)$ for $x, y : A$ are trivial.[243]

In this way, we get an h-stratification on all types, which for $h \geq 0$ corresponds to the stratification by homotopy types. Types A at level $h = 0$ are called sets. Why is that? By definition, in this case for any two terms $a, b : A$ the identity types $\mathrm{Id}_A(a, b)$ are mere propositions. In the topological interpretation, all path spaces $\mathrm{Id}_A(a, b)$ are thus either empty or point-like, i.e., there are no non-trivial paths or other higher homotopic information. The collection of all connected components thus forms a topological 0-type, i.e., a disjoint union of point-like spaces, which can be interpreted as a set. The types **Bool** and $\mathbb{N}$ are sets in this sense.

The types at level $h = 1$ can be understood as groupoids and not as categories, as might be suspected. In the topological interpretation, this means that there are non-trivial paths, but paths between paths are trivial. The type **Grp** of all groups is a genuine groupoid at level $h = 1$, as groups can have different isomorphisms, and is therefore not a set. The type $\mathbb{S}^1$ is also of level $h = 1$ and corresponds to the homotopy type of the circle $S^1 = \{(x, y) \in \mathbb{R}^2 \mid x^2 + y^2 = 1\}$. However, it is different as a type, as it is not a set in the above sense.

The higher types at level $h \geq 2$ are given by h-groupoids, which represent topological h-homotopy types.

Isomorphism and Equivalence

The concept of isomorphism in mathematics is extremely important and omnipresent. Two objects A and B in a given category are isomorphic if there are morphisms, i.e., structure-preserving arrows

$$f : A \longrightarrow B$$

and

$$g : B \longrightarrow A,$$

such that the compositions $f \circ g$ and $g \circ f$ coincide with the identity mappings id_B and id_A. If this is the case, we call f and g isomorphisms.

To give a simple example, consider groups with 2 elements within the category of groups. There is only one of these up to isomorphism. It is customary to write this group either as the additive group

$$\mathbb{Z}/2\mathbb{Z} = \{0, 1\} \text{ with } 1 + 1 = 0$$

or as the multiplicative group

$$\mu_2 = \{\pm 1\} \text{ with } (-1) \cdot (-1) = +1.$$

The isomorphism

$$f : \mathbb{Z}/2\mathbb{Z} \longrightarrow \mu_2$$

is given by the assignment $f(0) = +1$ and $f(1) = -1$.

Even though this example seems simple, many constructions in modern mathematics involve complications with isomorphisms. There, isomorphic objects should be identified as much as possible without concrete identifications being known or unique. A good example of this is the homeomorphism problem in topology. The question of whether two topological spaces X and Y are isomorphic, or equivalently homeomorphic, is an undecidable problem according to a theorem by Markov. Even if X and Y are isomorphic, a concrete isomorphism is usually not easy to specify. On the other hand, the additional information contained in existing isomorphisms is often very important, as we have seen with the infinity groupoids. When studying such things, it becomes clear that the concept of equality has a deep philosophical meaning. In Martin–Löf type theory, this is expressed in the existence of identity types.

In type theory, for two given types A, B in a universe U, the isomorphism type

$$\mathrm{Iso}(A, B) \text{ or } (A \cong B)$$

and the equivalence type

$$\text{Eq}(A, B) \text{ or } (A \simeq B)$$

are defined. The isomorphism type $\text{Iso}(A, B)$ is—as we did at the beginning of this chapter for categories—defined by the existence of inverse mappings $f: A \longrightarrow B$ and $g: B \longrightarrow A$. The natural mapping

$$\text{Id}_U(A, B) \longrightarrow \text{Iso}(A, B)$$

between identity type and isomorphism type is in the case of sets (in the sense of type theory) an isomorphism.[244] In type theory, a mapping $f: A \longrightarrow B$ is referred to as an equivalence, denoted by $A \simeq B$, if all homotopy fibres

$$\text{hfiber}(f, b) = \sum_{a:A} \text{Id}_B(f(a), b)$$

of f are contractible for each b. This corresponds topologically to a weak homotopy equivalence and defines the equivalence type

$$\text{Eq}(A, B) = \sum_{f:A \to B} \prod_{b:B} \text{iscontr}(\text{hfiber}(f, b)).$$

An equivalence $f: A \longrightarrow B$ can also be characterised by the existence of right- and left-inverse mappings $g, h: B \longrightarrow A$, such that the types

$$\text{Id}_{(A \to A)}(h \circ f, \text{id}_A) \text{ and } \text{Id}_{(B \to B)}(f \circ g, \text{id}_B)$$

are both inhabited.

Univalence

Vladimir Voevodsky has further developed the type theory of Martin–Löf with its central identity types. When considering simplicial sets as models of type theory, he discovered the univalence axiom.[245] This axiom is not a rule built into type theory like the rest, but an axiom that must be imposed afterwards. It does not hold in all models of type theory and its significance is currently not entirely clear. Nevertheless, we will try to explain the univalence axiom.

For two types A, B in a universe U, there is always a mapping[246]

$$v: \text{Id}_U(A, B) \longrightarrow \text{Eq}(A, B).$$

The univalence axiom states that the mapping v itself is an equivalence in a higher universe U, i.e.,

$$\text{Id}_U(A, B) \simeq \text{Eq}(A, B).$$

If A and B are sets in the sense of type theory, then the concept of equivalence coincides with the concept of isomorphism of sets.

For each predicate P of type $(U \longrightarrow \text{Bool})$, then with respect to equivalence, one has a variant of Leibniz's invariance rule

$$\frac{A \simeq B \quad P(A)}{P(B)},$$

i.e., the property P is preserved when replacing equivalent types. This follows from the univalence axiom, as this invariance rule for identity follows from path induction and under the mapping

$$v \colon \text{Id}_U(A, B) \longrightarrow \text{Eq}(A, B)$$

is still preserved.[247] Topologically, this principle is trivial, because the assignment $A \mapsto P(A)$ takes discrete values $\bot$ and $\top$ and is continuous. From this even follows the seemingly stronger statement

$$A \simeq B \implies A = B,$$

which results from the inverse v^{-1} of the mapping v (up to homotopy). This can be interpreted as equivalences being able to be upgraded to identities.

The hierarchy of identity, isomorphism and equivalence types provides room for some open questions. For example, it is not clear whether these three terms are exhaustive, or whether there are other useful equivalences. An important question is how the univalence axiom is to be formulated in a categorical semantic interpretation of type theory and for which semantics it even applies.[248]

Semantics and Reality

9

In today's societies, the process of truth finding is fraught with difficulties. The truth of scientific theses must be constantly questioned, because we only get closer to it in individual steps and even the scientific operation is not error-free. In the non-scientific area, complex or conditional statements are difficult to convey and some insights do not reach the public at all or slowly. In addition, there are different schools of thought, prejudices, influences on opinions and not least stupidity and ignorance.

Leibniz's dream, to be able to determine and prove the truth of assertions in all areas, not only encounters such banal, but also systematic limits of undecidability and incompleteness. Truth finding is nevertheless unavoidable. The search for a philosophical concept of truth goes far before Leibniz. The quality of this concept is described by Thomas Grundmann in reference to the correspondence theory in his book on epistemology:

> The concept of truth is strictly speaking not an epistemological concept. It picks out a relation between mind and world, namely the reference to something that is the case. It is therefore very similar to other basic semantic concepts such as reference or meaning. Although truth is a semantic concept, it plays a prominent role within epistemology.[249]

Over the centuries, an extremely rich literature has emerged, with partly controversial definitions of truth, in which truth bearers and truth makers are understood differently.[250] We will completely neglect most of these theories and focus on the language-analytical semantic truth theories.

The latter were introduced by Tarski in the 20th century with the help of formal languages. His method led to the concept of semantics, which he could define via the concept of a metalanguage that is richer than the given object language.[251] Suitable semantics illustrate mathematics and thus simplify understanding. In mathematics, set theory with the Zermelo–Fraenkel axioms or category theory is often used as a metalanguage. This gave rise to the field of model theory. However, this is only one of many options, as forms of type theory, category theory or set theory can be used both as object language and as metalanguage and are mutually interpretable. The question of consistency of a given theory does not become easier, because richer languages rather complicate this question than solve it. In this respect, semantic truth theories are closely related to the coherence theory of truth.

The concepts of equivalence and equality, which we encountered in type theory and in higher categories, allows a new structural view of the foundations of mathematics. The associated semantic concept of truth—in conjunction with the concept of equivalence—enables insights that are relevant with regard to natural and philosophical questions of science.

Finding the Truth

The pre-Socratic scholar Anaximander of Miletus lived around 600 BC. Although virtually nothing of his writings has survived, he is considered a significant natural scientist due to posthumous descriptions. Pliny the Elder wrote that Anaximander had opened the door to nature:

> Rerum fores aperuisse, Anaximander Miletus traditur primus.[252]

In one of his books, Carlo Rovelli vividly described how the development of natural sciences was influenced by Anaximander experienced a tremendous upswing because he replaced traditional flawed views with scientific ways of thinking. Rovelli writes:

> Anaximander opened the door to physics, geography, the study of meteorological phenomena, and biology. Beyond these important contributions, he set in motion the process that led to the rethinking of our worldview: our way of gaining knowledge, which is based on the rebellion against the obvious. In this sense, Anaximander can undoubtedly be called one of the fathers of scientific thinking.[253]

This was vividly documented in Anaximander's observations of celestial bodies. His logical arguments support the sphericity of the Earth's surface—as opposed to the Earth as a disc—and its free-floating position in space through observations and precise conclusions. Mathematics as a science had already developed outside the time of Anaximander earlier in Babylonian and Sumerian cultures and after this period within the influential school of Pythagoras.

Rovelli's example wonderfully illustrates the complexity of the concept of truth beyond Anaximander's considerations. Because when we say that the Earth's surface is a sphere, this statement is not literally true, not even the weakening that it is an ellipsoid. The poles of the Earth are flattened in different ways. If we look even more closely, we see that the Earth has an irregular surface and is filled with matter inside, which we do not fully understand in its innermost structure and whose composition has gaps in space. Therefore, the best form of this statement is that the Earth's surface is approximately spherical. The formulation by means of an approximation is not imprecise, because with a clever mathematical description, the deviation of the Earth's surface from an ideal sphere can be numerically limited by an inequality that a disc shape cannot fulfil. Thus, after a few steps, we have arrived at a formulation that excludes all claims that describe the Earth as a disc, which was the goal of all these considerations. Anaximander's arguments historically provided the first proof of this. In this example, we see very well that it can be difficult and sometimes requires several steps of thought to precisely grasp truth.

The Opposite of Truth

There is also the opposite of truth, untruth. Leibniz had already thought about the origin of untrue statements and attributed this to the lack or ignoring of evidence and other errors. He wrote about this in the "Nouveaux essais sur l'entendment humain" from 1704:

> After we have spoken sufficiently of all the means which let us recognise or suspect the truth, we want to say something about our errors and incorrect judgements. People must often be mistaken because there are so many disagreements among them. The causes of this can be traced back to the following four: 1) the lack of evidence, 2) the little skill to use it, 3) the lack of good will to make use of it, 4) the wrong rules of probability.[254]

By false probability rules, Leibniz understood unreliable guesses and estimates that people use in their judgements. Indeed, humans often fall prey to irrational prejudices.[255]

False statements can arise from errors that are oversights, if they are not consciously made. Closely related to the error is the concept of fallacy, which means a faulty implication in the sense of logic and in which the acting person feels justified. In colloquial language and the logic used there, fallacies are often made. Particularly noticeable is that the implication $A \Rightarrow B$ is often reversed without reason. Let's look at two examples and consider the implications:

I am currently eating a soup $\Longrightarrow$ I am alive

and

My boss does not appreciate my performance $\Longrightarrow$ My salary will not be increased.

Both implications are valid in general. This is independent of whether I am currently eating a soup or my boss really does not appreciate my performance, because an implication $A \Rightarrow B$ is always true, if the statement A is false. In the first implication, the reversal

I am alive $\Longrightarrow$ I am currently eating a soup

is usually nonsensical. However, this type of reversal is constantly applied in the second implication, even though there can obviously be many other reasons why salaries are not increased. Another fallacy is the linking of correlations with causations. A frequently cited example of this is the correlation between birth rate and the occurrence of storks in certain residential areas. It is obviously nonsensical to conclude the causality that babies are brought by storks. The truth is that there are other parameters that both promote or prevent, such as possibly the rural or urban location.

A state that goes beyond errors and fallacies and lasts longer we call a mistake. This is a constellation in the thought world of a person, in which the truth of certain statements is believed because the situation at the time of the mistake suggests apparent evidence.

Mistakes can arise in various ways. A special case are faulty or contradictory theories, in which any statement can be inferred. Russell's paradox, which appeared in Frege's work, can be considered the fundamental mistake of naive set theory. Only the Zermelo–Fraenkel axioms and Russell's type theory have shown a way out. In physics, Albert Einstein cleared up some of the mistakes of physics in his works in the annus mirabilis 1905, including the existence of the ether as the basic substance of the universe. Mistakes lead to new insights in this way.

A second and very threatening form of mistakes arises through misinformation, conspiracy theories, targeted manipulation of information or filter bubbles in communication media, which put people in the situation that they no longer question statements themselves. Resilience against such influences and the training of independent thinking are therefore indispensable.

The Truth Theory of Tarski and Kripke

Alfred Tarski wanted a formally correct theory of truth which further develops the Aristotelian correspondence theory and avoids paradoxes. In a famous work from 1933, he himself wrote:

> The present work is almost entirely devoted to a single problem, namely that of the definition of truth; its essence consists in the fact that one has to construct – with regard to this or that language – a factually accurate and formally correct definition of the term "true statement". This problem, which is counted among the classical questions of philosophy, causes significant difficulties.[256]

Tarski succeeded in this in the context of languages, which he called poorer languages. These include in particular the formal languages of logic and mathematics. He wrote about this:

> In the further course of the treatise, I will only consider those built up by scientific methods, which are known today, i.e., the formalised languages of the deductive sciences; their characterisation is given at the beginning of §2 ... With regard to the "poorer" languages, the problem of the definition of truth finds a positive solution.[257]

His main result Tarski formulated in this work as follows:

> For each formalised language of finite order, we can construct in the metalanguage a formally correct and factually accurate definition of the true statement, by using exclusively expressions of general-logical character, further expressions of the language itself, as well as terms from the field of morphology of the language, i.e., the names of the language expressions and the structural relations existing between them.[258]

In richer languages, which include the natural languages, this construction cannot be applied directly. Tarski's method can be interpreted in such a way that he constructed a semantics for certain poorer languages using a richer metalanguage, which describes the truth of statements. More concretely, Tarski used for his theory of truth a formal language L and a superordinate metalanguage M which contains L. He used for his new theory a truth predicate T, with which the truth of statements from L within M can be checked. The letter T stands for "true" or for Tarski.

Each truth predicate T must fulfil a famous adequacy condition of Tarski, which is colloquially formulated as follows:

> The sentence "Snow is white" is true if and only if snow is white.[259]

This equivalence is called Tarski's biconditional in the literature. Biconditionals of this kind are sometimes more abstractly expressed as

> The sentence "p" is true if and only if p

or in the formal way

$$T(p) \iff \tilde{p}.$$

Here, the left side is the application of the truth predicate T to p, or more precisely to a standard name for p, such as a Gödel number $\ulcorner p \urcorner$. The right-hand side consists of the translation $\tilde{p}$ of p into the metalanguage M. The biconditional is therefore a syntactic statement within M.[260]

Tarski demanded in his theory of truth such syntactic biconditionals as necessary conditions of adequacy. These appear at first glance like tautologies. Indeed, biconditionals are also called quotation deletions and have inspired the redundancy theory of truth, which—similar to Frege's comment on the prime number property of the number 5—considers the whole concept of truth to be dispensable.

For the actual definition of truth, Tarski used a large repertoire. As the associated metalanguage, he mostly used set theory with the Zermelo–Fraenkel axioms. The truth predicate T associated with an object language L is in this case given by the satisfiability of formulas in set-theoretical models. The truth of statements in L is verified in the richer language M by a proof. Another method of his is to reduce the truth of non-atomic statements which are linked by logical signs like $\wedge$ or $\vee$, by complete induction over the length of the concatenation to the truth of the atomic statements contained therein. A difficulty arose with statements that contain quantifiers and free variables at the same time and are neither true nor false in themselves. One way to circumvent this problem is the method of quantifier elimination.[261]

Tarski's theory avoids self-referential paradoxes, but cannot be applied to natural languages in which such are possible. Kripke made an important step in 1975 to extend the theory of truth to more general cases.[262] For this he used infinite hierarchies of formal languages

$$L_1 \subset L_2 \subset \cdots,$$

so that L_i each has a truth predicate T_i in L_{i+1} and limited himself to the fact that the T_i can only be partially defined. In this way, he managed to reach a fixed point in the limit of the languages L_i.

The Semantics of Arithmetic

The semantics of formal languages and deductive systems serves to give syntactic concepts a meaning in the Fregean sense. Semantics is traditionally mostly defined as a set-theoretical interpretation of a given deductive system, so that all inference rules remain fulfilled in the language of sets. The constant symbols, the function and relation symbols of the formal language are assigned with sets and mappings. For this, a base set A is used, so that the constants are elements of A, the n-ary functions f are interpreted by set-theoretic mappings

$$f : A^n \longrightarrow A$$

and the n-ary relations are interpreted as subsets of A^n. The resulting structure of sets and mappings is called a model. A formula φ is said to be true in this model if it is satisfiable under the interpretation.[263] Satisfiability is checked using the calculation rules of set theory, i.e., usually with the Zermelo–Fraenkel axioms and possible additional chosen axioms. The model relation

$$\models \varphi$$

expresses that the formula φ is satisfied in every conceivable model and can therefore be called universally valid.

An illustrative case is the complete Dedekind–Peano arithmetic with first-order predicate logic as the object language. The truth of arithmetic statements is defined by their satisfiability in the standard model of natural numbers. Andrzej Mostowski was able to show, using the transfinite hyperarithmetical hierarchy, that a truth predicate T in the second-order Dedekind–Peano arithmetic is syntactically formulable.[264]

The 4-squares theorem is an example of a concrete arithmetic statement of the form

$$\forall n \, \exists a \, \exists b \, \exists c \, \exists d \quad n = a^2 + b^2 + c^2 + d^2.$$

This statement is true, if—colloquially speaking—for every n a tuple (a, b, c, d) can be specified that satisfies the theorem. In the set-theoretic model, the satisfiability can be verified by the surjectivity of the mapping

$$\mathbb{N}^4 \longrightarrow \mathbb{N}, \ (a, b, c, d) \mapsto a^2 + b^2 + c^2 + d^2.$$

The infinity of the set

$$\mathbb{N} = \{0, 1, 2, 3, 4, \ldots\}$$

plays a significant role in this semantics. But where does this infinite set come from? The model by John von Neumann is

$$0 = \emptyset, 1 = \{\emptyset\}, 2 = \{\emptyset, \{\emptyset\}\}, \ldots, n + 1 = \{0, 1, \ldots, n\}, \ldots$$

An alternative is the definition

$$0 = \emptyset, 1 = \{\emptyset\}, 2 = \{\{\emptyset\}\}, \ldots, n + 1 = \{n\}, \ldots$$

In both cases, $\mathbb{N}$ results in complex sets. None of these possible definitions is to be preferred over another or particularly useful from a structural point of view.

Dedekind was aware of this ambiguity problem and he pursued a different approach. He considered an arbitrary set X with an initial element $*$ and an injective self-mapping

$$S \colon X \longrightarrow X,$$

such that $*$ is not in the image of S. This results in a set of natural numbers $\mathbb{N}$ within X by

$$0 = *, 1 = S(*), 2 = S(S(*)), \ldots$$

He then showed in a uniqueness theorem that all these avatars of the natural numbers are isomorphic to each other. For this, he used second-order predicate logic. In the first stage, this theorem does not indeed hold and there exist deviating non-standard models.

In older literature, interpretations are usually based on naive, material set theory. A somewhat more general and often more suitable approach is Henkin semantics in the form of an interpretation in category theory. To this day, set theory is the preferred basis for mathematical investigations and is considered the standard semantics of mathematics. It conceals intrinsic questions, as set theory is an axiomatic theory that itself has different models. The continuum hypothesis is an example of how the truth of a statement like

There exists a cardinal number κ with $\aleph_0 < \kappa < 2^{\aleph_0}$

can depend on the chosen model and can only be decided by additional axioms as the forcing construction shows.

Completeness and Model Theory

Leopold Löwenheim and Thoralf Skolem founded a mathematical theory of semantics in the 20th century, laying the foundation for model theory[265] and the works of Tarski and Gödel. The Löwenheim–Skolem theorem is named after them and was first proven by Löwenheim in 1915. Around 1920, Skolem provided a further, more general proof.

The Löwenheim–Skolem theorem states that the existence of an infinite model of a finitely or countably axiomatised mathematical theory, formulated in first-order predicate logic, implies for every (infinite) cardinal number κ a model of cardinality κ. This theorem, which is usually derived from the completeness theorem as a corollary, shows that deductive systems always have models of different cardinality and thus models in first-order logic are never unique. This is a remarkably noteworthy result.

As it turned out, Skolem's proof method was strong enough to derive Gödel's completeness theorem, which he proved in his dissertation in 1929. Gödel referred in his dissertation to the proof by Skolem:

> A similar method was used by Th. Skolem to prove the well-known theorem named after him and Löwenheim.[266]

The completeness theorem states that any statement formulated in a first-order axiomatic theory of predicate logic is provable if and only if it is universally valid, i.e., fulfilled in every set-theoretical model. In other words:

$$\models \varphi \text{ if and only if } \vdash \varphi.$$

It is clear that every provable statement is universally valid. Only the converse is the difficult part of this theorem. Modern proofs for the completeness theorem usually go back to the dissertation of Leon Henkin, who modified and generalised Gödel's proof. His proof also yielded a version of this theorem in type theory, i.e., in higher-level logic. The most important idea in the proof is to demonstrate the existence of sufficiently many suitable models, because only then is universality a sufficiently strong condition.[267]

The Löwenheim–Skolem theorem implies several seemingly paradoxical statements, such as the existence of countable non-standard models of the real numbers and countable or uncountable non-standard models of the natural numbers. A non-standard model of $\mathbb{N}$ can be constructed by a kind of adjunction. For this, we choose a new element c, which is assumed to be different from all natural numbers, and demand infinitely many conditions

$$c > n \text{ for each } n \in \mathbb{N}$$

which are expressible in first-order predicate logic. The Löwenheim–Skolem theorem, under this assumption, yields a new model with an infinitely large number c. Every model of $\mathbb{N}$ contains at least the numerals, i.e., the ordinary natural numbers of the form $n = S^n(0)$, which are created by n-fold application of the successor function S. Only the standard model of the numerals is unique in all models up to isomorphism.

Such statements and constructions do not contradict the uniqueness theorems for the natural and real numbers found in textbooks, as these are usually formulated in second-order predicate logic. For example, Dedekind showed the uniqueness of the natural numbers up to isomorphism using his recursion theorem. The Löwenheim–Skolem theorem even implies countable models of the real numbers. The uncountability of the real numbers in such countable models is a—in a certain sense—correct statement, which is called Skolem's paradox.[268]

These two theorems form the beginning of the field of model theory within mathematics. It investigates which set-theoretical interpretations exist for axiomatic mathematical structures and which statements about them follow from knowledge of the logical foundations. Among the most important results of this field, in addition to the mentioned theorems, is the compactness theorem, which was found by Kurt Gödel in his proof of the completeness theorem.[269]

The Undefinability Theorem of Truth

In the formal language of Dedekind–Peano arithmetic with first-order predicate logic, arithmetical sentences φ are in principle characterisable as true in the standard model by considering the truth predicate T as a function of the Gödel number of φ. A theorem by Tarski states that the subset of natural numbers which consists of the Gödel numbers of such true arithmetical sentences is not an arithmetical set. This implies that the Tarskian truth predicate T is not definable by a formula in the underlying formal language of first order. For this, a second- or higher-order arithmetic is needed, for which there are further assertions, the truth of which is not definable. In particular, the set of arithmetical sentences true in the standard model in the union of all levels is not decidable. It is not even recursively enumerable, due to the undecidability of Hilbert's 10th problem, because diophantine sets and thus recursively enumerable sets are always arithmetical. Tarski's theorem was known to Gödel and is often referred to as the undefinability theorem of truth. It should be noted that this result does not contradict Tarski's definition of truth. But it shows that strictly richer languages are needed to define a truth predicate. To prove this theorem, assume that the concept of truth is given by a class sign

$$\text{True}(x)$$

and apply Gödel's diagonal lemma as in the proof of the first incompleteness theorem to $\neg\text{True}(x)$ or colloquially

I am not true.

The fixed point then results in a contradiction.[270] From the correct mathematical proof of Tarski's theorem, it follows that the set of Gödel numbers of provable arithmetical sentences does not coincide with the set of Gödel numbers of sentences true in the standard model, from which a variant of Gödel's first incompleteness theorem follows.[271]

Gödel first thought about a semantic proof after his engagement with the completeness theorem and exchanged ideas about it with Tarski.[272] In such semantic versions of the incompleteness theorem, there is a Gödel sentence Q which is true in the standard model of natural numbers, but in the object language is neither provable nor refutable. In the non-standard models, such theorems are generally not true, because otherwise the completeness theorem would imply provability. The concept of truth thus corresponds to provability in the respective model.

It should be noted that the first incompleteness theorem is not the opposite of the completeness theorem. The naming of these two theorems is confusing. The sentence Q can be replaced by a specific theorem. A later found example for this is Goodstein's theorem, which holds in the standard model of natural numbers and can be proven by transfinite induction.[273]

Truth Theory of Non-classical Logic

There is also a concept of semantics for non-classical logics. In intuitionistic logic, the rules of the deductive system differ from classical logic, so that the concept of truth is closer to the concept of computability. In models, the satisfiability with intuitionistic logic is to be demonstrated. Since the time of Brouwer, there has been a close connection between intuitionism and topology. Tarski has constructed an interpretation of intuitionistic logic which uses the Heyting algebra **Off**(X) of open sets in a topological space X. In it, the negation operator

$$\neg U = (X \setminus U)^\circ$$

applied to an open set U is given by the open core $(X \setminus U)^\circ$ of the complement of U in X. With this definition it is generally not the case that $\neg\neg U = U$, unless U coincides with the open core of its closure in X, which is almost never the case. Thus, **Off**(X) is generally not a Boolean algebra.[274]

Even more general than intuitionistic logic is modal logic. This is very old and is connected with the theory of possible worlds by Leibniz. Possible worlds are semantic realisations in the form of different models. Modal logic was in parts already before Leibniz is known, for example, in Jean Buridan, a medieval philosopher. It contains, in addition to the two usual truth values true and false, possible and necessary statements:

$$\Diamond p: \text{ It is possible that } p$$
$$\Box p: \text{ It is necessary that } p.$$

This allows for statements that are called contingent because they are possible, but not necessary.

Ruth Barcan laid the foundations of modal predicate logic in her 1946 dissertation, building on the modal logical calculus of Clarence Irving Lewis. Before Lewis, Hugh MacColl had already established a modal logical system around 1900.[275] These systems consist of the calculus of classical logic and the modus ponens as a conclusion rule from the formula

$$\Box(A \Rightarrow B) \Rightarrow (\Box A \Rightarrow \Box B)$$

and the additional conclusion rule

$$\frac{A}{\Box A}.$$

All corresponding formulas for $\Diamond$ follow from the relationship

$$\Diamond A = \neg \Box \neg A.$$

Barcan discovered formulas, which are now known as Barcan formulas

$$\Diamond \exists x \, Fx \Rightarrow \exists x \Diamond Fx$$

and

$$\Box \forall x \, Fx \Rightarrow \forall x \Box Fx.$$

The second formula has gained slightly more acceptance than the first, as the first formula implies an existence statement in the real world for all objects to which the formula can be applied. In addition, there is the Buridan formula

$$\Diamond \forall x \, Fx \Rightarrow \forall x \Diamond Fx$$

and its reversal

$$\forall x \Diamond Fx \Rightarrow \Diamond \forall x \, Fx$$

which together imply the interchangeability of $\Diamond$ and $\forall$.

Saul A. Kripke introduced semantics for non-classical logics that generalise classical semantics. This Kripke semantics generalises the Tarskian semantics. At the age of 18, Kripke proved a completeness theorem for modal logic. Before him, Kurt Gödel, Alfred Tarski, Evert Willem Beth and Andrzej Grzegorczyk had studied the semantics of non-classical logic. These semantic interpretations were later developed further by André Joyal and others, leading to the concept of a categorical semantics which connects type theory (also called higher-order logic) with category theory and is called Kripke–Joyal semantics.[276]

Around Kripke and Barcan there was a long-lasting and still not finally clarified priority dispute over the introduction of some terms into philosophy, particularly about the ideas underlying the theory of names as treated in the book "Naming and necessity" by Kripke.[277]

Categorical Semantics of Type Theory

Leon Henkin has proven a completeness theorem for type theory, i.e., for higher-level logic.[278] In his proof, types are interpreted using objects and morphisms in **Set**, i.e., with sets and mappings, where not all objects and morphisms in **Set** appear as images. This results in subcategories of **Set**. This was mistakenly sometimes referred to as a disadvantage in parts of the literature and for this reason first-order logic was preferred. Such Henkin semantics are more suitable for considering mathematics structurally because they abstract away many technical aspects of type theories.

Voevodsky studied a model of homotopy type theory in the category **sSet** of simplicial sets to verify the univalence axiom.[279] More generally, homotopical categories provide the correct semantics for dependent Martin–Löf type theories. They allow techniques of homotopy theory within **Top** and **sSet** to be abstracted. Their objects correspond to topological spaces and they contain weak equivalences as a class of morphisms, which correspond to weak homotopy equivalences in topology.

Model categories are the most important representatives for homotopical categories. They were introduced by Daniel Quillen and his school and include, in addition to weak equivalences, two more classes of morphisms, called fibrations and cofibrations. Closely related are the path categories, introduced by Kenneth Brown as a variant of model categories. Both concepts are connected in that the fibred objects in model categories, i.e., the objects A, whose morphism $A \longrightarrow *$ to the terminal object is a fibration, form a path category. The structure-preserving equivalences of model categories and thus of homotopical categories, for example of **Top** and **sSet**, are given by so-called Quillen equivalences.

Homotopical categories are related to $(\infty, 1)$-categories. This has a deeper reason, as we are interested in the category of all homotopy types of topological spaces. Grothendieck had conjectured that the infinity groupoids like $\Pi_\infty(X)$ exactly classify the homotopy types of topological spaces. He called this the homotopy hypothesis. Under this hypothesis, the category ∞**Grpd** of all infinity groupoids is obviously an $(\infty, 1)$-category, because infinity groupoids are $(\infty, 0)$-categories.

With the help of Dwyer–Kan localisation, homotopical categories can be turned into $(\infty, 1)$-categories and higher elementary toposes. In the literature, they are usually not studied themselves, but their models in the form of weak Kan complexes and quasi-categories. Recently, attempts have been made to provide model-independent definitions.[280]

In the homotopical semantics for type theory, dependent types are interpreted as fibrations. This term is based on the corresponding topological concept. A morphism $p: E \longrightarrow B$ of topological spaces is called a (Hurewicz) fibration if it satisfies a certain lifting property of homotopies and thus of paths (see diagram).

In this interpretation, a fundamental problem arises because substitutions in type theory are interpreted as pullbacks of fibrations and in categories the concatenation of pullbacks is only unique up to isomorphism. With the help of a construction by Jean Bénabou, each fibration can be "split" in such a way that this annoying problem can be successfully tackled. Possible solutions have been proposed by Andrew Pitts with split-type categories, Peter Dybjer with categories with families, Martin Hofmann with categories with attributes, Steve Awodey with natural models, and Vladimir Voevodsky with C-systems. The latter are also syntactic categories or contextual categories.[281]

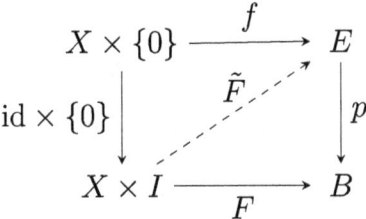

Lifting property of a Hurewicz fibration.

We want to explain path categories in more detail. In them, two types of morphisms are distinguished, called fibrations and weak equivalences, and in which path objects exist. Their properties are modelled after the topological case. There are thus some axioms that these types of morphisms must fulfil. For example, isomorphisms are always weak equivalences and the 2-out-of-3 rule applies: if two of the three morphisms $f, g, g \circ f$ are weak equivalences, so is the third. The composition of two fibrations and all isomorphisms are fibrations.

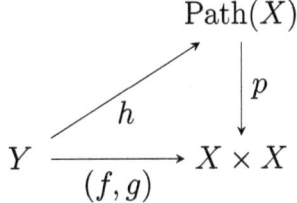

Homotopy with path objects.

Path categories possess path objects. This includes for each object X a path object $\mathrm{Path}(X)$ together with a composition of mappings

$$X \longrightarrow \mathrm{Path}(X) \longrightarrow X \times X,$$

so that the first mapping is a weak equivalence and the second mapping is a fibration. With the help of path objects, a concept of homotopy can be defined, in which two mappings $f, g \colon Y \longrightarrow X$ are homotopic if there is a lift h, as shown in the diagram.

Path objects exist in the case of topological spaces, because the path space $\mathrm{Path}_a(X)$ of paths with a fixed starting point a is contractible to the point path by reparametrisation and therefore the natural mapping

$$X \longrightarrow \mathrm{Path}(X)$$

for every topological space X is a homotopy equivalence. The projection mapping to the starting and ending point

$$p \colon \mathrm{Path}(X) \longrightarrow X \times X$$

is a fibration, in which all fibres correspond to the path spaces with a fixed starting and ending point.

The syntactic category $\mathcal{S}_T$ of a dependent type theory T is a path category under mild conditions.[282] The fibrations correspond to special projections of contexts. The path objects $\mathrm{Path}(A)$ are given by the contexts

$$x : A, y : A, p : \mathrm{Id}_A(x, y)$$

and the mappings

$$A \longrightarrow \mathrm{Path}(A) \longrightarrow A \times A$$

are given by $a \mapsto (a, a, id)$ and $(x, y, p) \mapsto (x, y)$. Every type theory T has a natural interpretation in its syntactic category $\mathcal{S}_T$. It can be shown that every other interpretation of T in a path category $\mathcal{C}$ can be factorised by the syntactic category and is therefore given by a functor of categories

$$\mathcal{S}_T \longrightarrow \mathcal{C}.$$

For this reason, it is interesting to further develop the syntactic categories, also known as contextual categories. Voevodsky began this study in the form of the C-systems he considered.

Categorical Logic

The assignment of the syntactic categories $\mathcal{S}_T$ to a type theory T has a reversal. Each infinity category or each higher elementary topos is based on a syntactic type theory called Mitchell–Bénabou language. It is equivalent to a higher-level logic, which maps the internal logic of arguing with the arrows in all diagrams of the category. It is the basis of categorical logic. We want to describe it in the case of an elementary topos.[283]

The types in the Mitchell–Bénabou language are formed from the objects of the elementary topos. The variables correspond to the identity morphisms

$$\mathrm{id} \colon A \longrightarrow A.$$

The terms of type B in n variables $x_1, \ldots, x_n$ arise from morphisms

$$f \colon X \longrightarrow B,$$

where $X = X_1 \times X_2 \times \cdots \times X_n$. Specifically, formulas φ are defined as terms of type

$$X \longrightarrow \Omega,$$

where Ω is the subobject classifier. With regard to the properties of Ω, φ defines a subobject of X within the elementary topos, which we briefly denote with

$$\{x \mid \varphi(x)\}.$$

Since Ω is a Heyting algebra, the logical operations in the Mitchell–Bénabou language are explained. The formula φ is fulfilled for an element

$$1 \xrightarrow{x_0} X,$$

if x_0 lies in the subobject $\{x \mid \varphi(x)\}$. This is illustrated in the depicted commutative diagram. Heyting algebras or Boolean algebras are thus crucial links between logic and elementary topoi.

Surprisingly, the Mitchell–Bénabou language does not necessarily satisfy the law of the excluded middle. Since a typical example of an elementary topos is the category **Sh**(X) of sheaves on a topological space X, this can be well read from the Heyting algebra **Off**(X) of the open sets in X, which we have already seen is usually not a Boolean algebra. The category **Set**, on the other hand, has classical logic with the law of the excluded middle as its internal logic.

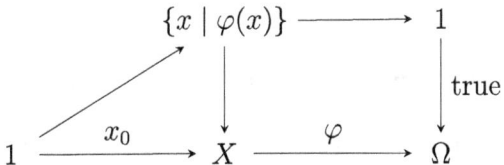

Validity of $\varphi(x_0)$.

The Mitchell–Bénabou language has a natural semantics in the underlying higher elementary topos. However, the validity of logical formulas there is sometimes difficult to grasp. The Kripke–Joyal semantics has proven to be a useful tool, which goes back to the works of Beth, Grzegorczyk, Kripke and Joyal on the semantics of non-classical logic. In the Kripke–Joyal semantics, the truth of formulas $\varphi(x)$ with

$$\varphi \colon X \longrightarrow \Omega$$

is defined on objects U that lie over X, i.e., for which a morphism $U \longrightarrow X$ exists. We say that the formula $\varphi(x)$ is true on U, in symbols

$$U \models \varphi(x),$$

if the depicted diagram commutes.

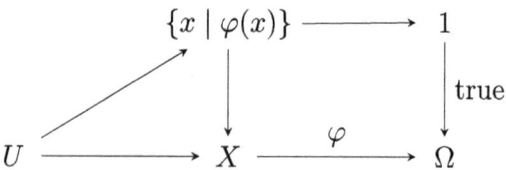

The forcing relation.

An object $U \longrightarrow X$ is called a generalised element of X, as it generalises points of the form $1 \xrightarrow{x_0} X$ that are not always readily available. The possibility that logical statements are only fulfilled on sets U is particularly important in connection with sheaf theory, as for example the existence of sections in sheaves on certain open subsets is a local property. The insight that truth can be local reveals a close connection between sheaf theory and logic. The quantifiers $\exists$ and $\forall$ can also be well described in this correspondence using adjoint functors. The relation

$$U \models \varphi(x)$$

is called the forcing relation, as the technique of forcing in set theory can be explained with the help of it in the context of suitable sheaves, as William Lawvere and Myles Tierney had recognised from about 1970 onwards. The original semantics of Kripke similarly emerges as a special case by considering sheaves on a suitable partially ordered set **P**.[284]

The Symbolic Construction of Reality

We initially discussed the correspondence and coherence theories of truth, which both seemed quite far apart to us. In the further course, we approached the coherence theory in a syntactic way through type theory. Alfred Tarski, on the other hand, has tried by his own account to further develop the correspondence theory of truth with the help of suitable semantics, which we found for type theory in categorical semantics. In the original formulations of the correspondence theory, however, a relationship with reality was demanded. How does this fit together?

Let's pause for a moment. The original correspondence theory is based on the assumption of a reality that raises many questions. Is there a reality at all, or is this a naive notion? What kind of existence are we talking about when it comes to nature and physical reality? How much of it can we perceive as humans? Is there a reality beyond physical reality, or is it always reducible to physical foundations, as reductionism postulates?

Physical objects have very abstract properties that are only sufficiently comprehensible with mathematical methods. Hermann Weyl has referred to the mathematical approach to understanding physics and other natural sciences as a symbolic construction of reality. In the introduction to an essay on this topic, he wrote in vivid and clear words, referring to Democritus:

It would probably not be a bad choice to date constructive natural sciences and critical philosophy from the day Democritus proclaimed: "Sweet and bitter, cold and warm as well as the colours, all this exists only in opinion, not in reality ($\nu\acute{o}\mu\omega$, $o\grave{\upsilon}$ $\varphi\acute{\upsilon}\sigma\epsilon\iota$); what really exists are unchangeable substance particles and their movement in empty space." Indeed, without questioning the standpoint of naive realism, there is no philosophy, and a theoretically constructive science of nature is impossible as long as one takes the phenomena as they are given with perception at face value.[285]

In the rest of the text, Weyl explains how science is shaped by the constructing spirit of man and how mathematics permeates the concepts of physics such as space and time. A well-known example of this is the modelling of spacetime by manifolds. We have questioned this description—following Riemann's habilitation thesis—as it is not clear whether spacetime can be modelled continuously on a small scale or whether a discrete mathematical model is more appropriate.

Weyl can be considered as the discoverer of gauge theories in physics. He translated the idea of covariance into the differential geometric concept of gauge theories. These are given by a Lie group G, so that the algebra $C^{\infty}(M, G)$ of G-valued differentiable functions on a spacetime manifold M operate on the physical gauge fields and the Lagrangian function $\mathcal{L}$ of the theory. Gauge fields are $\mathfrak{g}$-valued 1-forms A on M, where $\mathfrak{g}$ is the Lie algebra of G.[286] An element $g \in C^{\infty}(M, G)$ operates on the gauge fields by

$$A \mapsto g^{-1}Ag + g^{-1}dg.$$

In his first approach around 1919, Weyl dealt with a synthesis of electrodynamics and general relativity. Electrodynamics is a gauge theory with the gauge group $G = U(1)$, i.e., the complex numbers of absolute value 1 and the Lie group $\mathfrak{g} = i\mathbb{R}$. An elegant and simple formulation for this are the Maxwell equations in the compact notation

$$dF = 0$$
$$d * F = J.$$

Here, the field strength $F = dA$ is an exact 2-form, the magnetic vector potential A is a 1-form, the current J is a 3-form, $*$ is the Hodge $*$-operator and the gauge transformations are given by

$$A \mapsto A + d\varphi, \quad \varphi \text{ a scalar function.}$$

In a work from 1929 titled "Electron and gravity",[287] Weyl described the formalism of gauge theory even more precisely in a quantum mechanical context, by introducing a complex phase factor $\exp(i\varphi)$ that left the physics invariant. The gauge group G in this situation was also $U(1)$.

In more general cases, the gauge group G is not abelian. Such theories are called Yang–Mills theories, after a work by Chen N. Yang and Robert L. Mills from 1954.[288] The standard model of particle physics contains two gauge theories, the electroweak interaction with the gauge group $SU(2) \times U(1)$ and the strong interaction with

$G = SU(3)$. General relativity, i.e., gravitation, can be understood as a gauge theory with gauge group $G = SO(3, 1)$ according to Ryoyu Utiyama.

The operation of the gauge group on the gauge fields is analogous to homotopies of paths in topology. Therefore, it is useful to consider the gauge groupoid $\mathcal{G}(M)$ of all gauge fields. Similar to the fundamental groupoid Π_1, it carries more information than just the quotient space of the orbits of the gauge fields under the gauge group, as it also considers the automorphisms of each individual gauge field. This view has only established itself in recent years.

In addition to his considerations about the symbolic construction of reality, Weyl was interested in metaphysics and phenomenology.[289] He was particularly influenced by Kant, Fichte, Leibniz and Husserl. In metaphysics, attempts are made to conceptually approach the dimensions and limits of human existence. Such philosophical considerations, despite their large questions, which are extremely thought-provoking, also only provide access through language. Leibniz had already realised this and believed that man's gain in knowledge would be limited to a symbolic calculus and that the actual understanding of the world would be reserved for God alone.

Much suggests that for the description of reality in the natural sciences a symbolic calculus with suitable semantics is sufficient in the considered science, which makes a priori assumptions in the respective theory. Such considerations touch on the field of natural philosophy.[290] The deep, unresolved question of whether reality itself has a Platonic quality beyond the symbolic construction remains open. For it could be that our world, which most people imagine to be made purely of matter, is simply made up of facts that we can perceive at most in our minds through sensory experience or by recognising physical laws. Such sceptical, antirealist attitudes have a tradition dating back to Protagoras in antiquity and also appear in the immaterialism of George Berkeley.[291] They are the complete opposite of materialism, which only recognises the existence of the material world. Ludwig Wittgenstein coined the well-suited famous sentences in his "Tractatus logico-philosophicus":

> The world is everything that is the case.
> The world is the totality of facts, not of things.[292]

Equivalence and Truth Scepticism

Based on our considerations about the nature of reality, there are at least two corresponding ways to consider the concept of truth. The correspondence theory of truth in its original version makes sense in the experimental natural sciences, where the existence of a reality is usually not doubted and the results of physical and other experiments can at first glance be evaluated as a correspondence between theory and reality. On the other hand, there are sceptical, antirealist views, which regard reality and our perception of it only as a variant of a semantics. In such cases, as well as in mathematics, a coherence theory of truth is more applicable.

We must acknowledge that the concept of truth is not seen equally in every scientific discipline and it may be impossible to have a uniform concept of truth that represents a minimal consensus. This observation has given rise to newer philosophical attitudes that view the concept of truth sceptically.[293]

The diverse positions of truth relativism assume that truth is always dependent on a context or an evaluation perspective and cannot be absolute. Some of these views have the advantage of avoiding self-referentialities and are suitable for solving the difficulties in finding truth in natural languages and thus explaining phenomena such as faultless disagreements or defining the validity of future statements. They therefore require interesting further developments of the concept of semantics.

On the other hand, provably true statements are prototypes for absolute truths and form an important prerequisite for science and for the success of our coexistence. Only with larger changes in science, such as Kuhn's[294] paradigm shifts, are truths possibly corrected. For example, the laws of Newtonian physics were modified by the general theory of relativity. In this process, the description of physical space changed from flat Euclidean geometry to curved manifolds, in which Euclidean spaces are only contained infinitesimally in the form of tangent spaces.

An extreme position of truth relativism asserts that there can be no absolute truths in science because there are fundamentally no ultimate justifications and therefore all forms of truth are only relative. The possibility of Kuhnian paradigm shifts plays a rather controversial role here, as they seemingly describe a disruptive transition. However, Thomas Kuhn has defended himself against claims that his theory supports truth relativism.

How should such claims be evaluated? Vittorio Hösle emphasised that this form of truth relativism is affected by an antinomy, as the lack of an ultimate justification can be applied to truth relativism itself and thus leads to a performative self-contradiction.[295] This alone is already a significant counter-argument. Further criticism is directed against tendencies of arbitrariness, which result from the ethical consequences of such a theory due to insufficient societal consensus.

What about in science? Mathematical truths, especially theorems, are based on certain assumptions, denoted by Γ, and assert statements A. They are usually in all calculi of the form of a judgement

$$\Gamma \vdash A.$$

The assumptions in Γ can contain axioms. Through a correct proof of such a theorem, the absolute truth of the judgement $\Gamma \vdash A$ is shown or alternatively the relative truth of the statement A under the premise Γ. We have seen further absolute mathematical truths in the discussion of the Gödel–Lucas–Penrose argument.

In summary, it can be said that in mathematics there are both relative and absolute truths, which—given the constant premise Γ and in the same deductive system— apply under all conceivable circumstances and in all possible worlds. In any case, there are absolute truths in mathematics that are associated with the concept of provability.

With theories of physics, it is not much different. Sometimes it is said, that truth in the natural sciences is fundamentally based on measurements and agreement with nature. This ignores the hypothesis- and theory-building and falls short. Physical theories depend on underlying assumptions, like mathematical theories on their axioms.

Kuhnian paradigm shifts change these and enable new insights, but derivations within the old theory remain valid in the classical limit. Similar to mathematics, there are therefore relative and absolute truths in physics.

It is worth revisiting the example of the spherical shape of the Earth's surface. Using principles of equivalence or invariance,—similar to our discussion about the existence of ideal circles—neither an infinitely long descending chain of ultimate justifications nor a Kuhnian paradigm shift will ever change the approximate spherical shape of the Earth' surface. Such truths are called stable or absolute. Already Peirce pointed out the necessity of correcting insights through constant doubt, falsifications and the reorientation of theories when finding absolute truths in science.[296]

In generalisation of this example, there is a convincing approach to refute truth relativism by classifying truths with the help of the concept of equivalence in a stable or invariant sense. The idea of such an antirelativistic criterion of objectivity for truth goes back to Hermann Weyl, who liked to use the concept of equivalence and saw himself in this respect in the tradition of Felix Klein.[297]

How can Weyl's idea be formulated in the language that we have developed in this book? The truth of a statement φ in an object language L is described by interpretations $L \longrightarrow M$ in a suitable semantic metalanguage M in which φ can be proven. It has to be shown that—as in the example of the spherical shape of the Earth's surface—every conceivable deconstruction of truth is based on a refined interpretation $L' \longrightarrow M'$, so that the statement φ corresponds to an equivalent statement φ' in L'. The iterative process of deconstruction terminates with this strategy in appropriate equivalence classes of statements, which reinforces the meaningfulness of the demand for an absoluteness of truth. It is noteworthy that the concept of equivalence itself in a certain way represents a deconstruction of the concept of identity.

The Circle Closes

Which questions have we answered? The concepts of equivalence and univalence in type theory explain how equivalent objects relate to each other, without necessarily identifying them, as Frege had tried with the natural numbers. They acquire a logical character, as equivalences become exactly the transformations of abstract mathematical objects that preserve all relevant structures. This is related to the non-uniqueness problem and therefore with the Platonic world of ideas and is seen as the epitome of structural thinking. This entire edifice of thought is connected with ideas from Carnap, Grothendieck and Tarski on the invariance of logical truths.[298]

The concept of truth for type theory has a natural place in categorical semantics. But this is only one possibility, because type theory, category theory and set theory can be mutually interpreted in the form of object languages or metalanguages.[299] Through this perspective, the concepts of truth and semantics in mathematics are somewhat demystified, as they mean nothing more and nothing less than provability in richer deductive systems.

The circular image shows the three foundations set theory, type theory and (higher) category theory and their essential properties in an overall view. On the outer ring they are depicted as equals. The second ring describes the relationship of the respective theory in relation to syntactic or semantic aspects, because this—despite the possibility of mutual interpretations—plays a role. In the innermost ring, corresponding basic objects are found. In the table shown, some relationships between the three foundations are further specified.[300]

What questions remain open? The consistency question in mathematics is due to Gödel's incompleteness theorem in all three foundations in principle unsolved. Only through additional axioms of transfinite nature can the consistency of smaller parts of mathematics be secured. This approach shifts the consistency problem into other deductive systems at the price of further incompleteness.

What conditions must be placed on $(\infty, 1)$-categories to allow interpretations of type-theoretical substitutions and to be suitable from the standpoint of computability is also still an open question. Overall, the axiomatics of higher categories with their applications seem to be a promising research subject. This raises the fundamental question of whether (higher) category theory is independent of set theory and to what extent it represents a complete foundation of mathematics, as predicted by William Lawvere and others.

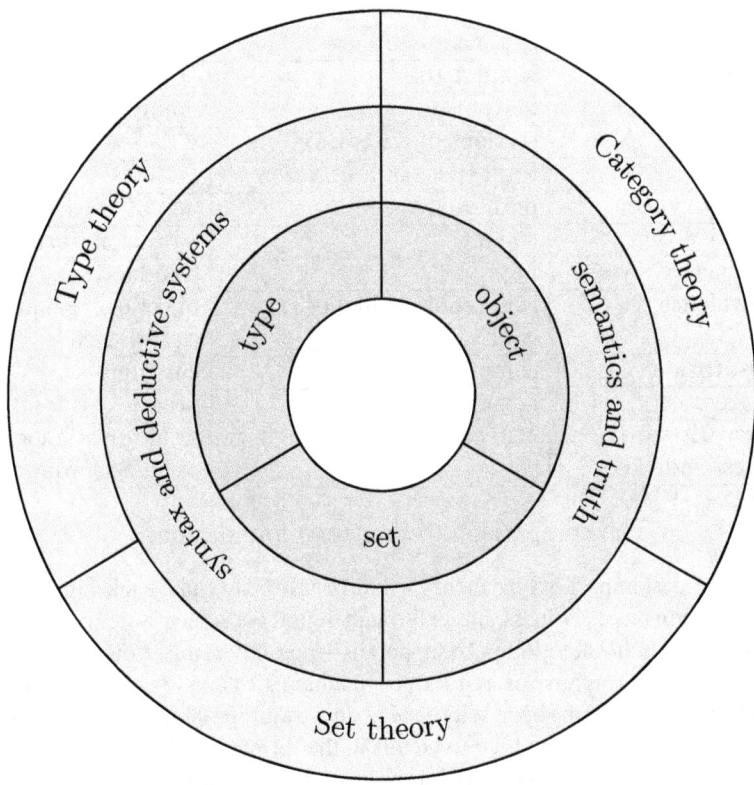

The three foundations of mathematics.

In type theory, it is open whether the axiom of univalence is the right choice and whether other developments—such as cohesive variants—are sensible and necessary. In connection with the interpretation of type theory in higher category theory, there is the opportunity to convey the syntactic and semantic perspective in university education, as has been practiced in computer science for some time.

Type theory	Category theory	Set theory, logic and topology
type	object	set, space
term $a : A$	$1 \xrightarrow{a} A$	element $a \in A$
universe	object classifier	universe
type $\mathbf{1}$	terminal object	$\top$, true
type $\mathbf{0}$	initial object	$\bot$, false
type of propositions	subobject classifier Ω	truth value set
power type	$\mathrm{Hom}(-, \Omega)$	power set
product type	product	$\wedge$
sum type	coproduct	$\vee$
function type	exponential object	$\Rightarrow$
$(A \to \mathbf{0})$	$\mathrm{Hom}(A, 0)$	$\neg A$
dependent type	morphism	family, fibration
$\prod B(x)$	sections of a fibration	$\forall$
$\sum B(x)$	total space of a fibration	$\exists$
identity type	path object	$=$, path space
inductive type	colimit	induction, recursion
coinductive Type	limit	coinduction
equivalence	isomorphism, higher equivalence	bijection, homotopy-equivalence
substitution	composition	cut rule
program	generalised element	proof
Curry–Howard correspondence	Mitchell–Bénabou language	intuitionistic logic of higher order

The comparison of the three foundations.

The strongest impact of type theory will probably be to change teaching, research and publication practices in the future through digital assistance systems for the verification of proofs and algorithms, to support us—possibly using artificial intelligence techniques—in the creative search for new mathematics and to achieve more transparency in the spirit of the open science idea. In a way, this approach is similar to that of the Bourbaki group, who tried to do this in the form of texts.[301] The possibilities that arise from such developments can be seen as a partial realisation of Leibniz's dream of a Lingua universalis.

References and Notes for Further Study

1. See [W60].
2. See [E93, Ch. 15].
3. See §17 of Leibniz's "Monadology" at www.projekt-gutenberg.org.
4. See [C36, T36].
5. See [T50].
6. Although Tarski claims to want to further develop the correspondence theory of truth, his approach is of a different nature, see [T35, T69]. We cannot fully represent the extensive literature on the concept of truth. See, however, [H96, JN19, S17].
7. See [G03, Vol. I, 1931].
8. See [G83, M84].
9. Logical concepts are part of mathematics.
10. See the neuroscientific research by Stanislas Dehaene.
11. Contained in [C32, D30, F84, F93].
12. See [F14] on the philosophy of jazz.
13. Contained in [F62].
14. See the articles on sense and meaning in [JN19, S17].
15. See [L, Ser. VI, Vol. 4A, No. 178] and [L17].
16. See [L, Ser. VII, Mathesis, No. 57].
17. See [DR07, J19] for Leibniz's ideas in general.
18. See the articles on being in [JN19, S17].
19. See [H57].
20. See [H57].
21. See [H57].
22. See [F94, T12].
23. See [L90, Book II, Ch. 27].
24. See [H39, Book I, Sec. 6].
25. The book [BP11] describes the lively goings-on in this salon. Famous is an anecdote, after which Hume made the remark on his first arrival in the salon that he did not know any atheists and d'Holbach replied in essence that he

could name him 15 of the 18 present immediately and 3 more who were not quite decided yet.

26. See [B98,C11] on the Platonic world of ideas.
27. René Descartes referred to such mutual relationships as dualisms. They play a role in the body-soul and in the mind-matter problem.
28. There are further intermediate forms like conceptualism, which was represented by William of Ockham. For immaterialism see [B10].
29. See [L17,F80] for nominalism. In recent times, Hartry Field, Paul Benacerraf and John Searle held nominalistic views.
30. If we reject the Platonic world of ideas and at the same time in nominalism only acknowledge the world of ideas in the form of human or divine thoughts, then we obtain a pantheistic ontological proof of God, because the laws of nature were already present at the Big Bang, when no humans existed.
31. See the article on universals in [JN19].
32. All proofs of God are—unsurprisingly—based on such axioms. See [G03, Vol. III, 1970].
33. See [L, Ser. VI, Vol. 4A, No. 8].
34. See the article on the question of realism in [S17, Ch. IV].
35. Kant, Frege and Russell had doubts whether existence is a predicate or rather a property of properties. See Frege's "Dialogue with Pünjer on existence" in [F83, p. 60] as well as the articles on existence in [JN19,S17].
36. See [B65,L17] on Benacerraf's dilemma.
37. See [B51].
38. See [CN37].
39. Contained in [D30].
40. Contained in [C32].
41. Dedekind used second-order predicate logic, so non-standard models of $\mathbb{N}$ could not occur in his work.
42. See [F76].
43. The axiom of choice is equivalent to the well-ordering theorem and Zorn's lemma. See [M10, Ch. II] for the foundations of set theory.
44. See [F84,F93].
45. See [F84, §46].
46. See [F84, §72].
47. See [F84, §72].
48. Printed in [F76].
49. Reprinted in [F76].
50. See [P02, Part One, Ch. 1, §VI]. The translation is by Elisabeth Küssner–Lindemann from the digitally available German edition of 1904 by Teubner Publishing House. See also [S70, p. 151].
51. This includes Crispin Wright, Charles Parsons and Bob Hale. For Hume's principle see [L17, Ch. 9].
52. On logicism and type theory see [LS86,RW10].
53. See [E93].
54. See [C01].

55. Accessible at www.projekt-gutenberg.org.

56. See [E93].

57. See [L, Ser. VI, Vol. 1, No. 8].

58. See the shorter version "Ars brevis" [L01], the "Ars magna" and the work "Logica nova" [L85].

59. See [H55].

60. The original and translation can be found in [B96,CL03]. Louis Couturat was a significant Leibniz researcher.

61. See [W68].

62. Here Johann Joachim Becher, George Dalgarno, Athanasius Kircher and Philippe Labbé are to be mentioned, see [B96,E93]. Interlinguistic planned languages were developed by Johann Martin Schreyer (Volapük), Giuseppe Peano (Latino sine flexione), Louis Couturat (Ido) and Ludwik Zamenhof (Esperanto).

63. See [J19] for Leibniz's scientific ideas.

64. See [L, Ser. I]. German translation in [L89, Vol. V-2].

65. See the essays in [L, Ser. VII, Vol. 4A].

66. See the letter to Christiaan Huygens from April 1691 [L, Ser. III, Vol. 5, No. 17].

67. The law states that for any statement A, either A or $\neg A$ is true. For historical remarks on this, see [LS86, Ch. II]. Only indirect proofs are known for the Thue–Siegel–Roth theorem and König's lemma.

68. See [J19, Ch. 8] and the text "Generales inquisitiones de analysi notionum et veritatum" from 1686 in [L, Ser. VI, Vol. 4A, No. 165].

69. See [B47,dM47,F79,P57]. Also contributing were John Venn, William Stanley Jevons, Charles S. Peirce, Christine Ladd–Franklin and Ernst Schröder's book "Algebra of logic" [S90].

70. See [F83, p. 139].

71. See [F79].

72. See [M10, Ch. I/II] for such concept formations.

73. See the beginning of [F79].

74. A lattice has a partial order structure $\leq$ and two operations $\wedge$ (or $\cap$) and $\vee$ (or $\cup$), analogous to intersection (infimum) and union (supremum). The partial order is connected with the two operations. For example, $A \leq B$ holds exactly when $A = A \wedge B$. In addition to the usual associative and commutative laws, the two absorption laws $A \vee (A \wedge B) = A$ and $A \wedge (A \vee B) = A$ also apply. Both operations are only semigroups, therefore a lattice is not a ring in general. Using the symmetric difference instead of $\vee$ yields a ring though, in fact a $\mathbb{Z}/2\mathbb{Z}$-algebra.

75. See [MM92, Ch. I, §8] on Boolean algebras and Heyting algebras.

76. See [A09].

77. See [vA86, Art. 1]. Thomas Aquinas systematically linked Aristotelian philosophy with Christian theology.

78. See [A09, 1051b].

79. See [H51, Sec. 4.11].

80. See the article on realism and antirealism in [S17, Ch. IV] as well as [HP21, P06] on phenomena of physics.
81. See [H96, T35, T69, JN19, S17] on theories of truth.
82. See [L, Ser. VI, Vol. 4A, No. 8].
83. See §33 of the "Monadology", available at www.projekt-gutenberg.org, as well as the further essays in [L, Ser. VI, Vol. 4A].
84. See [M94].
85. See [L17, Ch. 1].
86. See [L17, p. 15–16].
87. Quine doubted in his essay [Q51] not only Kant's distinction, but also the logical empiricism of the Vienna Circle.
88. See the SWR radio contribution [V13].
89. See [F62, p. 31].
90. See [F83, p. 189].
91. See [F03, p. 39].
92. See [F03, p. 50].
93. See [F84, §61].
94. See [F83, p. 60] and [E21].
95. Also called the liar's paradox. It probably goes back to Eubulides of Miletus, see [B15, Part II].
96. See [B15, Part I].
97. See [T35, T69].
98. See [B15, Part III] and [T69].
99. See [G03, Vol. I, 1931] and [T69].
100. See [H40].
101. See [D24].
102. See [S76] for the history of ancient mathematics.
103. Polynomials and algebraic varieties appear in mathematical models of almost all natural and social sciences. Two beautiful examples of this are quantum field theory and algebraic statistics.
104. In the Python programming language:
     ```
     def gcd(a, b):
         while b>0:
             a, b = b, a%b
         return a
     ```
105. Assuming there are only finitely many prime numbers $p_1 < p_2 < \cdots < p_n$, consider the number $N = p_1 p_2 \cdots p_n + 1$ and show that none of the prime factors of N appear in the finite list. Contradiction.
106. First show $F_m - 2 = 2^{2^m} - 1 = (2^{2^{m-1}} + 1)(2^{2^{m-1}} - 1) = F_{m-1}(F_{m-1} - 2)$ and then apply complete induction.
107. For a proof see www.mersenne.org and [MP11, §12].
108. Contained in [R90].
109. The Riemann–Siegel formula is contained in [S66].
110. Langlands programme addresses such questions.
111. Printed in [MS23].

112. See [MP11, §13].
113. A set with an order relation $\leq$ is totally ordered if for any two elements a, b, either $a \leq b$ or $b \leq a$ holds.
114. See [D93] for definitions and the well-ordering on trees.
115. An uncountable cardinal number κ is strongly inaccessible if it cannot be represented by cardinal arithmetic from smaller cardinal numbers. In particular, for every cardinal number $\alpha < \kappa$ it still holds that $2^\alpha < \kappa$. The Morse–Kelley axioms also imply the consistency of the Zermelo–Fraenkel axioms.
116. See [C63,C64] and [H18, Ch. 7]. The forcing method is inspired by Gödel's completeness theorem and the Löwenheim–Skolem theorem. A modern proof of Cohen's theorem comes from Lawvere and Tierney. It can be found in [MM92, Ch. VI].
117. See [S70, p. 149].
118. A monoid is a set M together with an associative operation $*: M \times M \longrightarrow M$ and a neutral element $e \in M$. It is free, when there is a subset $E \subseteq M$, such that every element of M can be uniquely written as a combination of finitely many elements of E.
119. A (multiplicative) group G is free, when there is a subset $E \subseteq G$, such that every element of G can be uniquely written as a combination of finitely many elements of E and their inverses.
120. A topological space is a set X, in which a collection of open subsets U is distinguished. Here, X and the empty set $\emptyset$ are open and (arbitrary) unions and finite intersections of open sets are open.
121. Due to the general theory of relativity, space is curved, so this statement is only approximately correct.
122. See [S76] for the history of non-euclidean geometry.
123. Mathematically, this is defined so that the pre-images of open sets are again open. In a metric space, this property is defined by the well-known ε-δ-definition.
124. Contained in [N83].
125. Simplicial sets are contained in the class **CGHaus** of compactly generated Hausdorff spaces, which include CW-complexes and locally compact Hausdorff spaces. See [GJ09, Ch. I] on simplicial sets.
126. See [MS25] and the collected works [N83].
127. A chain complex is a sequence $\cdots \longrightarrow C_{n+1} \longrightarrow C_n \longrightarrow C_{n-1} \longrightarrow \cdots$ of abelian groups C_i (or more generally of modules over a ring R) together with $\mathbb{Z}$-linear mappings $d_i: C_i \longrightarrow C_{i-1}$, which satisfy $d_{i-1} \circ d_i = 0$.
128. See [S76] for the history of mathematics. The book [RT30] refers to the historical significance of arithmetic and geometry.
129. See [W19] on these topics.
130. See [MP11, §5] for details on some protocols. In the ElGamal protocol, a suitable subgroup $G \subseteq U_p$ with q elements is often used, where q is a Sophie–Germain prime, i.e., $p = 2q + 1$ is also a prime number. The mathematician Sophie Germain dealt with Fermat's conjecture for such prime numbers as exponents.
131. Fermat's little theorem states that modulo a prime number p, always $g^p \equiv g$ for all g. From this follows $g^{p-1} \equiv 1$ for all g coprime to p.

132. The value of the Euler φ-function $\varphi(N)$ at the point N gives the number of numbers coprime to N between 1 and $N-1$, i.e., the order of U_N.
133. See [S97, W19].
134. The unpublished work is available at www.bitcoin.org.
135. See [S97].
136. See [HP21, P06] for covariance and spacetime.
137. Printed in [L89, Vol. V-2].
138. See [E16].
139. See [P06, Ch. 19].
140. Contained in [R90].
141. See [C95, HP21] and [P06, Ch. 32].
142. See www.claymath.org.
143. See [H49, R58] for Hebb's rule and the perceptron, [G16, K23] for deep learning, [V13] for learning algorithms and [W19] for the theoretical computer science behind all this.
144. See [MP43].
145. See [K23] for the mathematics of deep learning.
146. See [BD19].
147. See [J17].
148. Contained in [D30], see [MS23].
149. See [S99].
150. See [MS23, MS25].
151. See [R84] for all aspects of recursive functions and their hierarchies.
152. See [T77].
153. See [T36].
154. See [C40, T36, P36] and the book [K52].
155. In the Python programming language, we search for the solution $(3, 2)$ as follows:

```
k, n = 1, 2
while k**2–(n–k)**3! = 1:
    if k<n–1:
        k = k + 1
    else:
        k, n = 1, n + 1
print(k, n)
```

156. See [W19]. Concrete goals are realised by optimisation, sorting, and search algorithms. Well-known sorting algorithms are Quicksort and Bubblesort. Another class of algorithms are cryptographic algorithms and protocols, such as the symmetric protocols AES (advanced encryption standard) and DES (digital encryption standard) or the asymmetric public-key protocols like RSA cryptography, ECC (elliptic curve cryptography) and ElGamal cryptography.
157. See [W19] for algorithms and complexity theory.
158. See [R84, Ch. 5].
159. There are other characterisations of recursive functions in **P** through predicative recursion, see [BC92] and the literature therein.

160. The effort is $O(h(n))$, if there is a function h and a positive constant C such that for large values n the effort grows like $C \cdot h(n)$.
161. See [W19, Ch. 3] for **NP**-completeness.
162. See [C36, T36].
163. See [B94, D64].
164. See [M10, Ch. V].
165. Each relation r is given by an equation $u = v$, where u, v are words in the generating elements. In the case of groups, $v = 1$ without loss of generality.
166. See [D64, MS24, P47, T77].
167. See [T50].
168. See [W50].
169. See [B37, F79, MS23, P57, W22].
170. Significant examples of axiomatic theories appeared already at the end of the 19th century, like Euclidean geometry, whose axioms were closely examined by Moritz Pasch and David Hilbert, Zermelo–Fraenkel set theory and Dedekind–Peano arithmetic. See [L17] for formalism.
171. See [B47, M10, T13].
172. See [G35].
173. See [P45].
174. See [G03, Vol. I, 1931]. In fact, Gödel did not use Dedekind–Peano arithmetic, but the formal system in "Principia mathematica" [RW10]. For his proof, fragments of the Dedekind–Peano arithmetic are sufficient if they meet the Hilbert–Bernays–Löb criteria.
175. For a modern presentation see [H17] and [H18, Ch. 4].
176. See [S13] and [M10, Ch. II].
177. See [C34].
178. See [M94].
179. John von Neumann's letters to Gödel can be found in [G03, Vol. V]. Gödel's reply letters are partly only preserved in the form of handwritten drafts. See [HB39, P20, SS20] for further aspects.
180. See [HB34], the lectures from 1930 and 1931 in [ES13, App. D] as well as the correspondence between Gödel and Bernays in [G03, Vol. IV]. More about Gödel's theorems can be found in [C21, S13, T13].
181. See [G03, Vol. I, 1931].
182. See [A19], [C21, §5], [G35] and [S17].
183. See [T50, S92].
184. See [G03, Vol. III, 1951].
185. See [F62, T39].
186. See [L61, P89, P94, D93, F96].
187. See the appendix in [B85] for the history of these ideas.
188. See [BM84, B94, F78, SS20] and the appendix in [B85] on constructivism and intuitionism.
189. See [G03, Vol. I, 1933e].
190. See [MP11, §9] for a constructive proof.
191. Reprinted in [F76].

192. See [B85, Ch. IV].
193. The proof implements a weak form of the axiom of choice. A constructive proof outside of special cases is not known.
194. See [BM84, B94, F78].
195. Contained in [D30, MS23].
196. On Emmy Noether and her influence see [MS25].
197. See [C15, DK80, G15, L09, Q67].
198. On open subsets $U \subset X$ there are commutative algebras $\mathcal{O}_X(U)$ of holomorphic functions, which together form the structure sheaf $\mathcal{O}_X$. An analogous algebraic example are affine spectra $X = \mathrm{Spec}(A)$ for commutative rings A, which consist of all prime ideals in A and are endowed with the Zariski topology. By gluing together such affine spectra, the concept of the scheme by Alexander Grothendieck is obtained. Schemes also carry a structure sheaf $\mathcal{O}_X$. Instead of the Zariski topology, there are other topologies on a scheme such as the étale topology and other Grothendieck topologies, where the open sets are usually not subsets of X but morphisms $U \longrightarrow X$ in a category. Variants of schemes in non-archimedean geometry are the rigid-analytic spaces by John Tate and the perfectoid spaces by Peter Scholze, which open up new possibilities for research. Scholze received the Fields Medal in 2018 for his ideas.
199. The sheaf property refers to coverings. If $\mathcal{F}$ is a presheaf, U is an open set in X and $f \in \mathcal{F}(U)$ is an element, then open coverings $U = \bigcup_i U_i$ are considered. If we denote by f_i the restriction of f to U_i, then obviously the restrictions of f_i and f_j to the subset $U_i \cap U_j$ coincide for all pairs of indices i, j. A presheaf $\mathcal{F}$ is a sheaf if, conversely, given functions $f_i \in \mathcal{F}(U_i)$ with coinciding restrictions on $U_i \cap U_j$, a section $f \in \mathcal{F}(U)$ can be constructed, which delivers the function f_i on each U_i. The sheaf property is often symbolised by the exactness of the following equaliser sequence:

$$\mathcal{F}(U) \to \prod_i \mathcal{F}(U_i) \rightrightarrows \prod_{i,j} \mathcal{F}(U_i \cap U_j).$$

200. To define the stalk $\mathcal{F}_x$, the limit of all $\mathcal{F}(U)$ with $x \in U$ must be considered and it results in $\mathcal{F}_x$ as a set that depends on x. This limit object receives incoming arrows from all $\mathcal{F}(U)$ with $x \in U$ and is a universal object with this property, i.e., for every other object $\mathcal{G}$ with such incoming arrows there is an arrow from $\mathcal{F}_x$ to $\mathcal{G}$.
201. The reason is that there are injective boundary mappings $d_i : \Delta_{n-1} \longrightarrow \Delta_n$ on the one hand, which omit an element $0 \leq i \leq n$, and on the other hand surjective degeneration mappings $s_i : \Delta_n \longrightarrow \Delta_{n-1}$, which hit the element i twice. Each contravariant functor from Δ. to **Set** then provides a simplicial set S. in **sSet**.
202. If the collections $\mathrm{Hom}_{\mathcal{C}}(A, B)$ do not form sets, then $\hat{\mathcal{C}}$ can be replaced by the cocompletion of $\mathcal{C}$ which consists not only of presheaves.
203. See [L64].
204. See [MM92, Ch. IV, §8] and [LS86, Part II].
205. See [LS86, Part II].

206. See [MM92, Ch. I, §4] for an axiomatic definition of elementary toposes. In **Sh**(X), $\Omega(U) = \{V \mid V \subset U \text{ open}\}$.
207. For Grothendieck toposes see [I04].
208. Exponential objects exist in the category **CGHaus** of compactly generated Hausdorff spaces.
209. See [C15,G15,L08,L09].
210. See [RV22].
211. See [L08] for an easily accessible introduction.
212. They should be locally cartesian closed, i.e., every comma category $\mathcal{E}/X$ is cartesian closed. Furthermore, they should have a subobject classifier and an object $\mathbb{N}$ of natural numbers. Limits and colimits exist and correspond in some way to homotopy limits and -colimits. For $(\infty, 1)$categories, the transition to an elementary $(\infty, 1)$topos is possible by modification. See [BM18,C15,L09].
213. See [G15].
214. See [G83].
215. Such $(\infty, 1)$ categories result from a Dwyer–Kan localisation after the weak equivalences in model categories or the related path categories. Here, certain— previously non-invertible—morphisms are inverted and they become equiv- alences. This generalises the Gabriel–Zisman localisation of categories and originates in the construction of fractions, i.e., the localisation of a ring. See [DK80] and [C15, Ch. 7].
216. Schemes are generalisations of algebraic varieties that locally look like prime ideal spectra $\mathrm{Spec}(R) = \{\mathfrak{p} \mid \mathfrak{p} \text{ is a prime ideal in } R\}$.
217. Algebraic spaces are generalisations of schemes and are special cases of stacks. They naturally arise as quotients of schemes U by identification via an equiva- lence relation $R \subseteq U \times U$, where the projection maps $R \longrightarrow U$ are each étale, i.e., in particular unbranched.
218. Stacks are general space concepts that exist in algebraic, topological and differ- entiable form. Grothendieck and Artin developed algebraic stacks, which gener- alise algebraic schemes. They are more suitable as moduli spaces for algebraic objects and for forming quotients than schemes. A stack $\mathcal{X}$ is a category that is fibred over a base category $\mathcal{B}$ in the form of a functor $\pi \colon \mathcal{X} \longrightarrow \mathcal{B}$, where the fibres are groupoids. In addition, so-called descent data must be satisfied, which state that the family π forms a 2sheaf (more precisely a $(2, 1)$sheaf) of groupoids over $\mathcal{B}$ in a suitable Grothendieck topology. In algebraic geometry, $\mathcal{B}$ is usually given by the category $\mathcal{B} = (\mathbf{Sch}/S)_{\text{fppf}}$ of schemes over a fixed base scheme S and equipped with the fppf-topology. A scheme X over S can be con- sidered as a stack, where $\mathcal{X} = \mathbf{Sch}/X$ and under π a scheme U/X is mapped to U/S. The fibre over U/S consists as a category of the objects $\mathrm{Hom}_S(U, X)$ with id $: U \to U$ as the only morphism. All fibres are thus groupoids. In topol- ogy, $\mathcal{B}$ is typically the category **Top** and in differential geometry a category of manifolds.
219. See [AM69,C20,F82].
220. See [YM54].
221. See [AC21,C20,MM92,S21].

222. Instead of Hausdorff space, the property sober is sufficient. The subobject classifier Ω satisfies $\Omega(U) = \{V \mid V \subset U \text{ open}\}$ and thus the union of all open sets in X can be reconstructed. If X is sober, this allows the determination of X, see [MM92, Ch. IX, §3].

223. See [HMS17, MSP13, S02].

224. See [C40, LS86, RW10].

225. See [M84, M94].

226. A property is intensional when the content or the essence play a role and extensional when it depends on the external scope.

227. See [V13].

228. See [LS86, Part III].

229. In addition to $\mathbb{N}$, $\mathbb{Z}$, $\mathbb{Q}$ and $\mathbb{R}$, the complex numbers $\mathbb{C}$, the quaternions $\mathbb{H}$ and the Cayley octaves $\mathbb{O}$.

230. Universes were originally an idea of Paul Bernays, see [G18]. For type-theoretical universes see [V13, Sec. 1.3].

231. See [HS98, KV91, L10].

232. See [LS86, Part II].

233. See [CH86].

234. See [B18].

235. [V13, App. A].

236. See [M10, Ch. I].

237. See [V13, Ch. 3] for the Curry–Howard correspondence. It is related to the Brouwer–Heyting–Kolmogorov and to the realizability interpretation.

238. In [V13, Ch. 1.11] the corresponding $p : P$ is constructed.

239. See [G08, S21] and the web addresses isabelle.in.tum.de, coq.inria.fr, wiki.portal.chalmers.se/agda, leanprover.github.io and the unimath library on github.com.

240. See [G03, Vol. II, 1958] and [R84, Thm. 4.8]. Gödel's idea originated in 1941. It goes back to Hilbert's unsuccessful attempt to prove the continuum hypothesis, see [H26].

241. See [AW09, HS98, V13].

242. The proof can be found in [V13, Rem. 3.11.2].

243. See [V13, Rem. 3.11.2].

244. See [A14, §5].

245. See the introduction of [V13].

246. See [V13, Sec. 2.10].

247. See [A14] for a proof.

248. The so-called Rezk completion turns a category into a univalent one.

249. See [G08, Ch. 2].

250. The correspondence theory has different variants in the literature. In addition, there are many attitudes of scepticism that reject aspects of reality or the concept of truth, or shift them to social consensus and speech acts. Descriptions of these theories can be found in [G08, H96, JN19, S17].

251. See [B15, H96, K59, K65, K75, K80, T35, T69, S17].

252. See [P, Book II].

253. See [R19, p. 12].
254. See [L, Ser. VI, Vol. 6]. German translation by Ernst Cassirer available at www. project-gutenberg.org.
255. Astonishing examples of this can be found in [R19].
256. See [T35], first appeared in 1933 in Polish language in [T33].
257. See [T35].
258. See [T35].
259. See [T35, T69].
260. See [B15, H96, T69].
261. See [T31] and the overview article [T69].
262. See [K75].
263. An atomic formula φ in n variables corresponds to a relation and is mapped by the interpretation to a subset of A^n. Then $\varphi(t_1, \ldots, t_n)$ is true if the interpretation of the vector $(t_1, \ldots, t_n)$ lies in this subset. For composite formulas, truth is defined by reducing to the atomic components. See [M10, Ch. II, Sect. 2].
264. See [M51].
265. See [M10, Ch. X] on model theory.
266. See [G03, Vol. I, 1930].
267. See [H49, H50, M10]. The proof by contradiction of the theorem (in classical logic) assumes that φ is universally valid, but not provable. Then the addition of $\neg\varphi$ remains consistent. By adjoining free symbols for constants to the underlying formal language, a model is formed in which $\neg\varphi$ is valid. This is a contradiction, since φ was valid. A topos-theoretical proof goes back to Pierre Deligne. For further aspects see [A21] and [MM92, Ch. IX].
268. Skolem rejected the existence of uncountably infinite sets.
269. The compactness theorem states that every (potentially infinite) set of formulas has a model if and only if every finite subset has models.
270. See [M10, p. 256] and [S13].
271. See [M10, Ch. II, §11].
272. See [P20].
273. See [H18, Ch. 4.5].
274. See [MM92, Ch. VI] and the original work [T38].
275. See [B46, M06].
276. See [B56, G64, K65, T38] on intuitionistic logic, [K59] on modal logic and [G03, Vol. I, p. 299].
277. See [K80]. Many people were involved in this dispute.
278. See [H50]. See [LS86, Part II] for a different proof. In these generalisations, the compactness theorem and the Löwenheim-Skolem theorem are abandoned in their original formulation.
279. See [V13, Ch. 4].
280. For model categories see [Q67]. The Dwyer–Kan localisation is also called simplicial localisation. It localises with respect to weak equivalences and higher morphisms arise. See [DK80]. For $(\infty, 1)$-categories and higher elementary toposes see [BM18, C15, J02, L09, Q67]. Model-independent definitions are studied in [RV22].

281. See [AL18], [B85], [C86], [H94] and [S91].
282. See [B18].
283. See [MM92, Ch. VI].
284. See [K65, T72] and [MM92, Ch. VI].
285. See [W68, Vol. IV, p. 289].
286. The setup includes the group G, a principal fibre bundle P on M with group G and the associated vector bundle E under the adjoint representation. The gauge fields A are connections on E and the Lagrangian function $\mathcal{L}$ includes the curvature F of the respective connection A.
287. See [W68].
288. See [YM54].
289. The name metaphysics goes back to [A09], phenomenology to Edmund Husserl.
290. See the essays on natural philosophy in [S17, Ch. V]. Kurt Gödel was of the opinion that physics, like mathematics, was synthetic a priori in its theory assumptions, see [G03, Vol. III, p. 360].
291. See [B10] and [S17, Ch. IV].
292. See [W22]. Wittgenstein can be seen as a precursor of poststructuralism in his later works.
293. This includes truth pluralism, which traces back to Crispin Wright and Michael Lynch, as well as the truth relativism of John MacFarlane and others. The latter is also widespread in poststructuralist philosophies and includes the idea of deconstruction. For mathematical aspects, see [T12].
294. See [K62].
295. See [H90]. Theodor Adorno and Karl Popper also rejected truth relativism for other reasons.
296. See [P78] and the article on pragmatism in [JN19].
297. See [D88]. Hermann Weyl saw open problems in the application of his ideas to quantum mechanics, as the measurement process introduces discontinuities. Even from today's perspective, the understanding of the measurement process in quantum mechanics still appears incomplete. Attempts to explain it through non-linear theories with hidden variables were not successful.
298. See [C28, A17, A18]. It would be interesting to revisit Hume's principle in this context.
299. See [A11].
300. See "Relation between type theory and category theory" on ncatlab.org.
301. Similar open science projects are MathOverflow, nLab, n-Category Café, Stacks Project and Stack Exchange.

Bibliography

[AL18]   Ahrens, B., Lumsdaine, P.L.F.: Vladimir Voevodsky: categorical structures for type the-
          ory in univalent foundations. Log. Methods Comput. Sci. **14**(3), 1–16 (2018)
[AC21]   Anel, M., Catren, G.: New Spaces in Mathematics. Cambridge University Press (2021)
[vA86]   von Aquin, T.: Quaestiones disputatae de veritate, Quaestio I, Philosophische Bibliothek,
          Band 384, Felix Meiner Verlag (1986)
[A09]   Aristoteles: Metaphysik, Philosophische Bibliothek, Band 308, Felix Meiner Verlag
          (2009)
[A19]   Artemov, S.: The Provability of Consistency (2019). ArXiv:1902.07404
[AM69]   Artin, M., Mazur, B.: Étale Homotopy, vol. 100. Lecture Notes in Mathematics. Springer
          (1969)
[AW09]   Awodey, S., Warren, M.A.: Homotopy theoretic models of identity types. Math. Proc.
          Cambridge Philos. Soc. **146**, 45–55 (2009)
[A11]   Awodey, S.: From sets to types, to categories, to sets. In: Foundational Theories of
          Classical and Constructive Mathematics, pp. 113–125. Springer (2011)
[A14]   Awodey, S.: Structuralism, invariance, and univalence. Philos. Math. **22**, 1–11 (2014)
[A17]   Awodey, S.: Carnap and the invariance of logical truth. Synthese **194**, 65–78 (2017)
[A18]   Awodey, S.: Univalence as a principle of logic. Indag. Math. **29**(6), 1497–1510 (2018)
[A21]   Awodey, S.: Sheaf representations and duality in logic. Outstanding Contributions to
          Logic, vol. 20, pp. 39–57. Springer (2021)
[B98]   Balaguer, M.: Platonism and Antiplatonism in Mathematics. Oxford University Press
          (1998)
[B46]   Barcan, R.: A functional calculus of first order based on strict implication. J. Symb.
          Logic **11**(1), 1–16 (1946)
[B85]   Beeson, M.: Foundations of constructive mathematics, Ergebnisse der Mathematik und
          ihrer Grenzgebiete, 3. Folge, vol. 6, Springer (1985)
[BC92]   Bellantoni, S., Cook, S.: A new recursion theoretic characterization of the polytime
          functions. Comput. Complex. **2**(2), 97–110 (1992)
[B85]   Bénabou, J.: Fibred categories and the foundations of naive category theory. J. Symb.
          Log. **50**, 10–37 (1985)
[B65]   Benacerraf, P.: What numbers could not be. Philos. Rev. **74**, 47–73 (1965)

[BD19]   Ben-David, S., Hrubeš, P., Moran, S., Shpilka, A., Yehudayoff, A.: Learnability can be undecidable. Nat. Mach. Intell. **1**, 44–48 (2019)

[B18]   van den Berg, B.: Path categories and propositional identity types. ACM Trans. Comput. Log. **19**(2), 1–32 (2018)

[BM18]   van den Berg, B., Moerdijk, I.: Exact completion of path categories and algebraic set theory: part I. J. Pure Appl. Algebra **222**(10), 3137–3181 (2018)

[B10]   Berkeley, G.: A Treatise Concerning the Principles of Human Knowledge. Aaron Rhames Printer (1710)

[B56]   Beth, E.W.: Semantic construction of intuitionistic logic. Koninklijke Nederlandse Akademie van Wetenschappen **19**(11), 357–388 (1956)

[B96]   Blanke, D.: Leibniz und die Lingua Universalis, Sitzungsberichte der Leibniz-Sozietät, Band 13. Heft **5**, 27–35 (1996)

[BP11]   Blom, P., Philosophen, B.: Ein Salon in Paris und das vergessene Erbe der Aufklärung, Hanser Verlag (2011)

[B37]   Bolzano, B.: Wissenschaftslehre, Seidel (1837)

[B51]   Bolzano, B.: Paradoxien des Unendlichen, Reclam Verlag (1851)

[B47]   Boole, G.: The Mathematical Analysis of Logic: Being an Essay Towards a Calculus of Deductive Reasoning. Macmillan (1847)

[B15]   Brendel, E.: Die Wahrheit über den Lügner. De Gruyter Verlag (2015)

[B94]   Bridges, D.: Computability, Graduate Texts in Mathematics, vol. 146. Springer (1994)

[BM84]   Bridges, D., Mines, R.: What is constructive mathematics? Math. Intell. **6**(4), 32–38 (1984)

[B73]   Brown, K.S.: Abstract homotopy theory and generalized sheaf cohomology. Trans. Am. Math. Soc. **186**, 419–458 (1973)

[C32]   Cantor, G.: Gesammelte Werke. Springer (1932)

[C20]   Carchedi, D.J.: Higher Orbifolds and Deligne–Mumford Stacks as Structured Infinity Topoi, vol. 264. Memoirs of the American Mathematical Society (2020)

[C28]   Carnap, P.R.: Der logische Aufbau der Welt. Weltkreis Verlag (1928)

[C34]   Carnap, P.R.: Logische Syntax der Sprache. Springer (1934)

[C86]   Cartmell, J.: Generalised algebraic theories and contextual categories. Ann. Pure Appl. Logic **32**, 209–243 (1986)

[C11]   Cassou-Noguès, P.: On Gödel's "platonism." Philosophia Scientiae **15**(2), 137–171 (2011)

[CN37]   Cavaillès, J., Noether, E. (Hrsg.): Briefwechsel Cantor–Dedekind, Hermann Éditeurs (1937)

[C21]   Cheng, Y.: Current research on Gödel's incompleteness theorem. Bull. Symb. Log. **27**(2), 113–167 (2021)

[C36]   Church, A.: An unsolvable problem of elementary number theory. Am. J. Math. **58**, 345–363 (1936)

[C40]   Church, A.: A formulation of the simple theory of types. J. Symb. Logic **5**, 56–68 (1940)

[C15]   Cisinski, D.-C.: Catégories supérieurs et théorie des topos, Séminaire Bourbaki, 67ème année (2014–2015). Exposé No. **1097**, 1–57 (2015)

[C63]   Cohen, P.J.: The independence of the continuum hypothesis, part I. Proc. Natl. Acad. Sci. USA **50**(6), 1143–1148 (1963)

[C64]   Cohen, P.J.: The independence of the continuum hypothesis, part II. Proc. Natl. Acad. Sci. USA **50**(6), 105–110 (1964)

[C95]   Connes, A.: Non-Commutative Geometry. Academic Press (1995)

[CH86]   Coquand, T., Huet, G.: The calculus of constructions, INRIA rapport de recherche 530 (1986)

[C14]   Coquand, T.: Théorie des types dépendants et axiome d'univalence, Séminaire Bourbaki, 66ème année (2013–2014). Exposé No. **1085**, 1–18 (2014)

[C01]   Couturat, L.: La loqique de Leibniz, Félix Alcan Éditeur (1901)

[CL03]   Couturat, L., Leau, L.: Histoire de la langue universelle, Hachette Éditeurs (1903)

[D64]   Davis, M. (ed.): The Undecidable. Raven Press (1964)

[D93]   Davis, M.: How subtle is Gödel's theorem? More on Roger Penrose. Behav. Brain Sci.
          16, 611–612 (1993)
[D30]   Dedekind, R.: Gesammelte mathematische Werke. Vieweg Verlag (1930)
[D88]   Deppert, W., Hübner, K., Oberschelp, A., Weidemann, V. (Hrsg.): Exakte Wissenschaften
          und ihre philosophische Grundlegung, pp. 445–467. Peter Lang Verlag (1988)
[DR07]   De Risi, V.: Geometry and monadology. Birkhäuser Verlag, Leibniz's analysis situs and
          philosophy of space (2007)
[D93]   Dershowitz, N.: Trees, Ordinals and Termination. Lecture Notes in Computer Science,
          vol. 668, pp. 243–250. Springer (1993)
[D24]   Diophant: Arithmetica, nicht überliefert (um 250 AD)
[DK80]   Dwyer, W., Kan, D.: Simplicial localizations of categories. J. Pure Appl. Algebra **18**(1),
          17–35 (1980)
[E93]   Eco, U.: La ricerca della lingua perfetta nella cultura europea. Editori Laterza (1993)
[E21]   Edmonds, D.: Die Ermordung des Professor Schlick. Beck Verlag, C. H (2021)
[E16]   Einstein, A.: Die Grundlage der allgemeinen Relativitätstheorie. Ann. Phys. **49**, 769–822
          (1916)
[ES13]   Ewald, W., Sieg , W., (eds.): David Hilbert's Lectures on the Foundations of Arithmetic
          and Logic 1917–1933. Springer (2013)
[F62]   Feferman, S.: Transfinite recursive progressions of axiomatic theories. J. Symb. Log.
          27(3), 259–316 (1962)
[F78]   Feferman, S.: Constructive theories of functions and classes, pp. 159–224. Logic Collo-
          quium 1978, North Holland (1978)
[F96]   Feferman, S.: Penrose's Gödelian argument. Psyche **2**(7), 21–32 (1996)
[F14]   Feige, D.M.: Philosophie des Jazz. Suhrkamp Verlag (2014)
[F94]   Fichte, J.G.: Grundlage der gesamten Wissenschaftslehre. Gabler Verlag (1794)
[F80]   Field, H.: Science Without Numbers. Oxford University Press (1980)
[F79]   Frege, G.: Begriffsschrift. Louis Nebert Verlag (1879)
[F84]   Frege, G.: Die Grundlagen der Arithmetik. Koebner Verlag (1884)
[F93]   Frege, G.: Grundgesetze der Arithmetik, Pohle Verlag (1893)
[F62]   Frege, G.: Funktion. Begriff, Bedeutung, Vandenhoeck und Ruprecht (1962)
[F76]   Frege, G.: Wissenschaftlicher Briefwechsel. Felix Meiner Verlag (1976)
[F83]   Frege, G.: Nachgelassene Schriften, 2. Felix Meiner Verlag, Auflage (1983)
[F03]   Frege, G.: Logische Untersuchungen, 5. Vandenhoeck und Ruprecht Verlag, Auflage
          (2003)
[F82]   Friedlander, E.: Étale Homotopy of Simplicial Schemes. Princeton University Press,
          Annals of Mathematics Studies (1982)
[G35]   Gentzen, G.: Untersuchungen über das logische Schließen I. Math. Z. **39**, 176–210
          (1935)
[G03]   Gödel, K.: Collected Works, vols. I–V, Oxford University Press (1986–2003)
[GJ09]   Goerss, P.G., Jardine, J.F.: Simplicial Homotopy Theory. Birkhäuser Verlag (2009)
[G08]   Gonthier, G.: Formal proof-the four-color theorem. Not. Am. Math. Soc. **55**(11), 1382–
          1393 (2008)
[G16]   Goodfellow, I., Bengio, Y., Courville, A.: Deep Learning. MIT Press (2016)
[G18]   Grayson, D.: An introduction to univalent foundations for mathematicians. Bull. Am.
          Math. Soc. **55**, 427–450 (2018)
[G83]   Grothendieck, A.: A la poursuite des champs, unveröffentlichtes Manuskript (1983)
[G15]   Groth, M.: A short course on ∞-categories. In: Handbook of Homotopy Theory, pp.
          549–617. Chapman and Hall (2020)
[G08]   Grundmann, T.: Analytische Einführung in die Erkenntnistheorie. De Gruyter Verlag
          (2008)
[G64]   Grzegorczyk, A.: A philosophically plausible interpretation of intuitionistic logic. Indag.
          Math. **26**, 596–601 (1964)
[H96]   Halbach, V.: Axiomatische Wahrheitstheorien. Akademie Verlag (1996)
[H40]   Hardy, G.H.: A Mathematician's Apology. Cambridge University Press (1940)

[HP21]   Hawking, S., Penrose, R.: Was sind Raum und Zeit? Klett–Cotta Verlag (2021)
[H49]   Hebb, D.O.: The Organization of Behavior-A Neuropsychological Theory. Wiley (1949)
[H57]   Heidegger, M: Identität und Differenz. Klett–Cotta Verlag (1957)
[H49]   Henkin, L.A.: The completeness of the first-order functional calculus. J. Symb. Log.
          14(3), 159–166 (1949)
[H50]   Henkin, L.A.: Completeness in the theory of types. J. Symb. Log. **15**(2), 81–91 (1950)
[H26]   Hilbert, D.: Über das Unendliche. Math. Ann. **95**, 161–190 (1926)
[HB34]   Hilbert, D., Bernays, P.: Grundlagen der Mathematik I, Grundlehren der mathematischen
          Wissenschaften, Band 40. Springer (1934)
[HB39]   Hilbert, D., Bernays, P.: Grundlagen der Mathematik II, Grundlehren der mathematis-
          chen Wissenschaften, Band 41, Springer (1939)
[H51]   Hobbes, T.: Leviathan, Andrew Crooke (1651)
[H55]   Hobbes, T.: De corpore, Andrew Crooke (1655)
[H90]   Hösle, V.: Die Krise der Gegenwart und die Verantwortung der Philosophie. Beck Verlag,
          C. H (1990)
[H17]   Hoffmann, D.W.: Die Gödelschen Unvollständigkeitssätze, 2. Auflage, Springer Spek-
          trum (2017)
[H18]   Hoffmann, D.W.: Grenzen der Mathematik, 3. Auflage, Springer Spektrum (2018)
[H94]   Hofmann, M.: On the Interpretation of Type Theory in Locally Cartesian Closed Cate-
          gories. Lecture Notes in Computer Science, vol. 933, pp. 427–441. Springer (1994)
[HS98]   Hofmann, M., Streicher, T.: The groupoid interpretation of type theory. In: Twenty-Five
          Years of Constructive Type Theory (Venice, 1995), Oxford Logic Guides, vol. 36, pp.
          83–111 (1998)
[HMS17]   Huber, A., Müller–Stach, S.: Periods and Nori Motives, Ergebnisse der Mathematik und
          ihrer Grenzgebiete, 3. Folge, vol. 65, Springer (2017)
[H39]   Hume, D.: A Treatise of Human Nature. John Noon (1739)
[I04]   Illusie, L.: What is a topos? Not. Am. Math. Soc. **51**(9), 1060–1061 (2004)
[JN19]   Jordan, S., Nimtz, C.: Grundbegriffe der Philosophie, 2. Reclam Verlag, Auflage (2019)
[J17]   Jost, J.: Object oriented models vs. data analysis–is this the right alternative? Boston
          Studies in the Philosophy and History of Science, vol. 327, pp. 253–286. Springer (2017)
[J19]   Jost, J.: Leibniz und die moderne Naturwissenschaft. Springer (2019)
[J02]   Joyal, A.: Quasicategories and Kan complexes. J. Pure Appl. Algebra **175**(13), 207–222
          (2002)
[K81]   Kant, I.: Kritik der reinen Vernunft. Johann Friedrich Hartknoch Verlag (1781)
[KV91]   Kapranov, M., Voevodsky, V.: ∞-groupoids and homotopy types. Cahiers de topologie
          et géometrie differéntielle **32**(1), 29–46 (1991)
[K52]   Kleene, S.C.: Metamathematics. North Holland Publishing Company (1952)
[K59]   Kripke, S.A.: A completeness theorem in modal logic. J. Symb. Log. **24**(1), 1–14 (1959)
[K65]   Kripke, S.A.: Semantical analysis of intuitionistic logic I. Stud. Logic Found. Math. **40**,
          92–130 (1965)
[K75]   Kripke, S.A.: Outline of a theory of truth. J. Philos. **72**, 690–716 (1975)
[K80]   Kripke, S.A.: Naming and Necessity. Blackwell (1980)
[K62]   Kuhn, T.S.: The Structure of Scientific Revolutions. University of Chicago Press (1962)
[K23]   Kutyniok, G.: The Mathematics of Artificial Intelligence. In: Proceedings of the ICM
          2022, pp. 5118–5139. EMS Press (2023)
[LS86]   Lambek, J., Scott, P.J.: Introduction to Higher-Order Categorical Logic. Studies in
          Advanced Mathematics, vol. 7. Cambridge University Press (1988)
[L64]   Lawvere, W.: An elementary theory of the category of sets. Proc. Natl. Acad. Sci. USA
          52, 1506–1511 (1964)
[L]   Leibniz, G.W.: Akademieausgabe der Leibniz Edition. www.leibnizedition.de
[L89]   Leibniz, G.W.: Philosophische Schriften, Wissenschaftliche Buchgesellschaft (1989)
[L17]   Linnebo, Ø.: Philosopy of Mathematics. Princeton University Press (2017)
[L85]   Llull, R.: Die neue Logik, Philosophische Bibliothek, Band 379. Felix Meiner Verlag
          (1985)

[L01]   Llull, R.: Ars brevis, Philosophische Bibliothek, Band 518. Felix Meiner Verlag (2001)

[L90]   Locke, J.: An Essay Concerning Humane Understanding. Thomas Basset and Edward Mory (1690)

[L17]   Lolli, G.: Ambiguità, un viaggio fra letteratura e matematica, Il Mulino (2017)

[L61]   Lucas, J.R.: Minds, machines and Gödel. Philosophy **36**, 112–127 (1961)

[L10]   Lumsdaine, P.L.F.: Weak ω-categories from intensional type theory. Log. Methods Comput. Sci. **6**, 1–19 (2010)

[L08]   Lurie, J.: What is an ∞-category? Not. Am. Math. Soc. **55**(8), 949–950 (2008)

[L09]   Lurie, J.: Higher Topos Theory. Annals of Mathematical Studies, vol. 170. Princeton University Press (2009)

[M06]   MacColl, H.: Symbolic Logic and its Applications. Longmans (1906)

[MM92]   MacLane, S., Moerdijk, I.: Sheaves in Geometry and Logic. Springer Universitext (1992)

[M10]   Manin, Y.I.: A Course in Mathematical Logic for Mathematicians. Graduate Texts in Mathematics, vol. 53, , 2nd edn. Springer (2010)

[M84]   Martin–Löf, P.: Intuitionistic Type Theory. Bibliopolis (1984)

[M94]   Martin–Löf, P.: Analytic and Synthetic Judgements in Type Theory. In: Kant and Contemporary Epistemology, pp. 87–99. Kluwer Academic Publishers (1994)

[MP43]   McCulloch, W., Pitts, W.: A logical calculus of the ideas immanent in nervous activity. Bull. Math. Biophys. **5**, 115–133 (1943)

[dM47]   De Morgan, A.: Formal Logic. Taylor and Walton (1847)

[M51]   Mostowski, A.: A classification of logical systems. Studia Philosophica **4**, 237–274 (1951)

[MP11]   Müller-Stach, S., Piontkowski, J.: Elementare und algebraische Zahlentheorie, 2. Vieweg+Teubner Verlag, Auflage (2011)

[MSP13] Müller-Stach, S.: What is a period? Not. Am. Math. Soc. **61**(8), 898–899 (2013)

[MS23]   Müller-Stach, S.: Richard Dedekind: Was sind und was sollen die Zahlen?" und, Stetigkeit und Irrationalzahlen", 2. Springer Verlag, Auflage (2023)

[MS24]   Müller–Stach, S.: Max Dehn, Axel Thue, and the undecidable. In: Lorenat, J., McCleary, J., Remmert, V., Rowe, D., Senechal, M. (eds.) Max Dehn: Polyphonic Portrait, History of Mathematics, vol. 46. American Mathematical Society (2024)

[MS25]   Müller–Stach, S.: Emmy Noether und ihre Bedeutung für die moderne Mathematik. In: Wie kommt das Neue in die Welt? (Mechthild Koreuber Hrsg.). Springer (2025)

[N83]   Noether, E.: Gesammelte Abhandlungen. Springer (1983)

[P57]   Peano, G.: Opere Scelte. Edizioni Cremonese (1957)

[P78]   Peirce, C.S.: How to make our ideas clear. Pop. Sci. Monthly **12**, 286–302 (1878)

[P89]   Penrose, R.: The Emperor's New Mind. Oxford University Press (1989)

[P94]   Penrose, R.: Shadows of the Mind. Oxford University Press (1994)

[P06]   Penrose, R.: A Road to Reality: A Complete Guide to the Laws of the Universe. Vintage Books (2006)

[P20]   von Plato, J.: Can Mathematics be Proved Consistent? Sources and Studies in the History of Mathematics and Physical Sciences. Springer (2020)

[P]   Plinius der Ältere: Naturalis historia (77 A.D.) (2024)

[P02]   Poincaré, H.: La science et l'hypothése. Éditions Flammarion (1902)

[P45]   Pólya, G.: How to Solve it. Princeton University Press (1945)

[P36]   Post, E.: Finite combinatory processes. J. Symb. Log. **1**, 103–105 (1936)

[P47]   Post, E.: Recursive unsolvability of a problem of Thue. J. Symb. Log. **12**, 1–11 (1947)

[Q67]   Quillen, D.: Homotopical Algebra. Lecture Notes in Mathematics, vol. 43. Springer (1967)

[Q51]   Quine, W.V.O.: The two dogmas of empiricism. Philos. Rev. **60**, 20–43 (1951)

[RT30]   Rademacher, H., Toeplitz, O.: Von Zahlen und Figuren. Springer Verlag (1930)

[RV22]   Riehl, E., Verity, D.: Elements of ∞-Category Theory. Studies in Advanced Mathematics, vol. 194. Cambridge University Press (2022)

[R90]   Riemann, B.: Gesammelte Werke. Springer (1990)

[R58]   Rosenblatt, F.: The perceptron-a probabilistic model for information storage and organization in the brain. Psychol. Rev. **65**(6), 386–408 (1958)

[R84]   Rose, H.E.: Subrecursion–Functions and Hierarchies, vol. 9. Oxford Logic Guides (1984)

[R19]   Rosling, H.: Factfulness. Ullstein Taschenbuch (2019)

[R19]   Rovelli, C.: Die Geburt der Wissenschaft–Aanximander und sein Erbe. Rowohlt Verlag (2019)

[RW10]   Russell, B., Whitehead, A.N.: Principia Mathematica. Cambridge University Press (1910)

[S21]   Scholze, P.: Liquid tensor experiment. Exp. Math. 1–6 (2021)

[S17]   Schrenk, M. (Hrsg.): Handbuch Metaphysik. Metzler Verlag (2017)

[S90]   Schröder, E.: Algebra der Logik, drei Bände. Teubner Verlag (1890–1905)

[SS20]   Schütte, K., Schwichtenberg, H.: Mathematische Logik. In: The Legacy of Kurt Schütte, pp. 71–91. Springer (2020)

[S92]   Searle, J.: The Rediscovery of the Mind. MIT Press (1992)

[S97]   Shor, P.W.: Polynomial-time algorithms for prime factorization and discrete logarithms on a quantum computer. SIAM J. Comput. 26, 1484–1509 (1997)

[S13]   Sieg, W.: Hilbert's Programs and Beyond. Oxford University Press (2013)

[S66]   Siegel, C.L.: Gesammelte Abhandlungen. Springer (1966)

[S99]   Siegelmann, H.T.: Neural Networks and Analog Computation. Birkhäuser Verlag (1999)

[S70]   Skolem, T.: Selected Works in Logic. Universitetsforlaget Oslo (1970)

[S13]   Smullyan, R.M.: Truth and provability. Math. Intell. 35(1), 21–24 (2013)

[S02]   Soulé, C.: The Work of Vladimir Voevodsky. In: Proceedings of the ICM 2002, vol. 1, pp. 99–103 (2002)

[S17]   Stępień, ŁT., Stępień, T.J.: On the consistency of the arithmetic system. J. Math. Syst. Sci. 7, 43–55 (2017)

[S91]   Streicher, T.: Semantics of Type Theory. Birkhäuser Verlag (1991)

[S76]   Struik, D.J.: Abriss einer Geschichte der Mathematik. Deutscher Verlag der Wissenschaften (1976)

[T13]   Takeuti, G.: Proof Theory, 2nd edn. Dover (2013)

[T13]   Tapp, C.: An den Grenzen des Endlichen. Springer (2013)

[T31]   Tarski, A.: Sur les ensembles définissables de nombres réels I. Fundam. Math. 17, 210–239 (1931)

[T33]   Tarski, A.: Pojęcie prawdy w językach nauk dedukcyjnych. Nakładem Towarzystwa Naukowego Warszawskiego (1933)

[T35]   Tarski, A.: Der Wahrheitsbegriff in den formalisierten Sprachen. Studia Philosophica 1, 261–405 (1935)

[T38]   Tarski, A.: Der Aussagenkalkül und die Topologie. Fundam. Math. 31, 103–134 (1938)

[T69]   Tarski, A.: Truth and proof. Sci. Am. 63–77 (1969)

[T12]   Tasić, V.: Poststructuralism and deconstruction: a mathematical history. Cosm. Hist. J. Natl. Soc. Philos. 8(1), 177–189 (2012)

[T77]   Thue, A.: Selected Mathematical Papers. Universitetsforlaget Oslo (1977)

[T72]   Tierney, M.: Sheaf Theory and the Continuum Hypothesis. Lecture Notes in Mathematics, vol. 274, pp. 13–42. Springer (1972)

[T36]   Turing, A.M.: On computable numbers with an application to the Entscheidungsproblem. Proc. Lond. Math. Soc. 42, 230–265 (1936)

[T39]   Turing, A.M.: Systems of logic based on ordinals. Proc. Lond. Math. Soc. 45(2), 161–228 (1939)

[T50]   Turing, A.M.: Computing Machinery and Intelligence, Mind, vol. LIX(236), pp. 433–460 (1950)

[V13]   Valiant, L.: Probably Approximately Correct. Basic Books (2013)

[V13]   Voevodsky, V., et al.: Homotopy Type Theory: Univalent Foundations of Mathematics. Princeton (2013)

[V13]   Vossenkuhl, W.: Frege–Das Problem der Wahrheit, "SWR2 Wissen: Aula" radio contribution (2013)

[W22]   Wardhaugh, B.: Begegnungen mit Euklid. Harper Collins (2022)

[W68]   Weyl, H.: Gesammelte Abhandlungen, Springer (1968)
[W50]   Wiener, N.: Human use of Human Beings–Cybernetics and Society. The Riverside Press
         (1950)
[W19]   Wigderson, A.: Mathematics and Computation. Princeton University Press (2019)
[W60]   Wigner, E.: The unreasonable effectiveness of mathematics in the natural sciences. Com-
         mun. Pure Appl. Math. **13**(1), 1–14 (1960)
[W68]   Wilkins, J.: An essay towards a real character und a philosophical language. Royal
         Society (1668)
[W22]   Wittgenstein, L.: Tractatus logico-philosophicus. Paul Kegan (1922)
[YM54]   Yang, C.N., Mills, R.L.: Conservation of isotopic spin and isotopic gauge invariance.
         Phys. Rev. **96**, 191–195 (1954)

Name and Subject Index